University Research for Industry

Topical Conference at the 2011 AIChE Spring Meeting and 7th Global Congress on Process Safety

Chicago, Illinois, USA
13 – 17 March 2011

ISBN: 978-1-61782-893-5

Printed from e-media with permission by:

Curran Associates, Inc.
57 Morehouse Lane
Red Hook, NY 12571

Printed by Curran Associates, Inc. (2011)

AIChE
3 Park Avenue
New York, NY 10016-5991

Phone: (203) 702-7660
Fax: (203) 775-5177

www.aiche.org

Additional copies of this publication are available from:

Curran Associates, Inc.
57 Morehouse Lane
Red Hook, NY 12571 USA
Phone: 845-758-0400
Fax: 845-758-2634
Email: curran@proceedings.com
Web: www.proceedings.com

TABLE OF CONTENTS

Prediction of Hydrogen Solubility in Heavy Hydrocarbons over a Range of Temperatures and Pressures Using Molecular Dynamics Simulations

Huajun Yuan [1], *Chris Gosling* [2], *Peter Kokayeff* [3] *and Sohail Murad* [1], *(1) University of Illinois at Chicago, Chicago, IL, (2) UOP LLC, Des Plaines, IL, (3) UOP LLC, Riverside, IL*

The solubility of gases in liquids can range over many orders of magnitude. Although fundamental principles of thermodynamics governing solubility are quite well understood, it is still not possible to make accurate predictions of solubility for many industrially important systems using thermodynamics principles only. Industrial projects are at the same time driving the need for increasingly complex and faster measurements with smaller samples under difficult conditions. It has been pointed out in particular, the need to properly account for the effect of the underlying fundamental interactions at the molecular level on the macroscopic behavior of fluids. Molecular-based studies of gas-fluids system can play an important role in meeting the physical property-related needs of the chemical industry, as a feasible alternative to experimental measurements, especially at high temperatures and high pressures. For example, the solubility of gases in liquids is an important problem in phase equilibrium. It is more complicated than conventional vapor–liquid equilibrium, since in many cases the solubility of gases in liquids can be very low (mole fractions of the order of 10^{-5}) at standard pressures, making accurate experimental measurements or theoretical predictions exceedingly difficult.

Possible solubility prediction applications include safe storage of organic liquids, recovery of chemical wastes, and property data estimation for process design. In particular, hydrogen solubility is needed for the modeling of hydrogenation processes, where such data is extremely crucial but traditional equation of state (EOS) methods do not provide accurate enough estimates for such systems. The solubility data is important to determine the optimal amount of hydrogen needed in the hydrogenation process and to improve plant safety. Excess hydrogen can result in energy being wasted and is also a potential safety hazard if too much is added during hydrogenation. Due to the complexity of heavy hydrocarbons (mixtures of various heavy components), experiments are particularly difficult to perform. In addition, hazards associated with extreme conditions in many industrial processes (high temperatures and/or high pressures), making extensive experimentation not an attractive option. Modeling is therefore an ideal and safe alternative, provided that the methods and parameters of molecular level modeling can be validated experimentally.

Using the method of molecular dynamics (MD) simulation, we have estimated the solubility of hydrogen in heavy hydrocarbons for a range of temperatures and pressures. Heavy hydrocarbon systems are known to be challenging not only for experimental measurements but also for reliable estimations using traditional equations of state. The simulation system used was designed with semi-permeable membranes to mimic actual experimental studies of gas solubility. Below is a typical simulation system, with the amount of hydrogen adjustable according to different initial hydrogen pressure as shown in Figure 1.

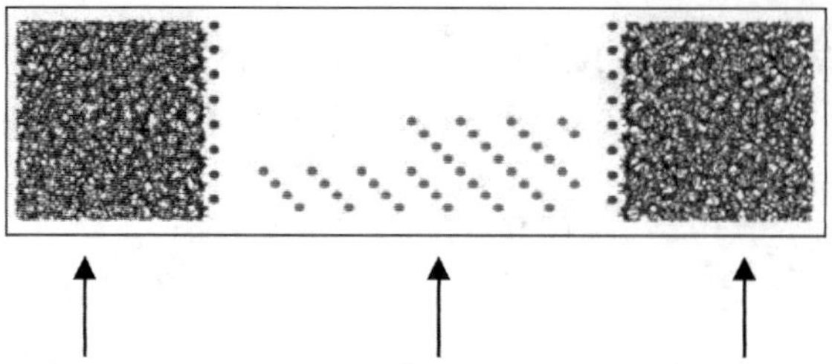

Fig. 1. Simulation system with hydrogen introduced, 480 Squalane + 128 Membrane + 40 H_2 molecules.

Binary interaction parameters between the solute gas and the solvent (heavy hydrocarbons) components were adjusted when necessary to improve agreement with experimental results and then used in subsequent multi-component studies. Figure 2 shows a typical Henry's constant estimation for squalane system at 300K.

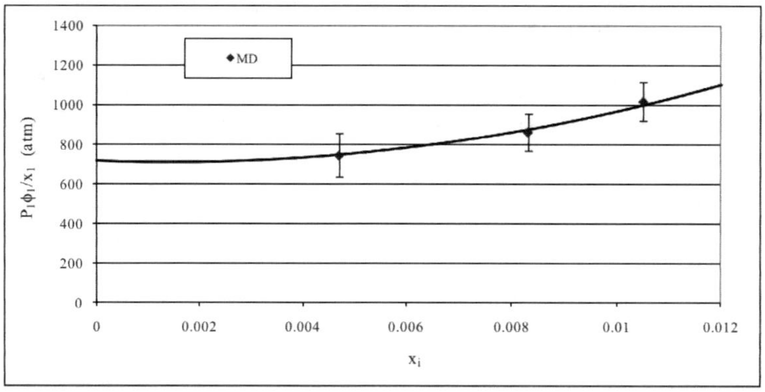

Fig. 2. Pressure dependence of solubility of H_2 in squalane, Henry's constant was obtained by extrapolating to $x_i \rightarrow 0$. The solid line is a polynomial fit of the data with a slope of 0 at $x_i = 0$.

After the interaction between hydrogen and heavy hydrocarbon component were determined, the heavy hydrocarbon system can then be represented using two different methods: (i) Functional group components method and (ii) Pseudo components method respectively. We can then predict the solubility of hydrogen in the mixed compounds at various temperatures and pressures. Temperature and pressure ranges studied included higher temperatures (up to 685 K) and pressures (up to 11.0 MPa) which are especially difficult to investigate experimentally.

Simulation results were finally used to adjust the binary interaction parameters (BIP) in simulation packages (e.g. ASPEN) to enable quick and reliable predictions. RK-Aspen (Redlich-Kwong-Aspen) equation of state is used as a suitable equation for hydrocarbon systems which can not only deal with nonpolar system, but also can be extended to polar-nonpolar system. An ASPEN pseudo component with a density and average boiling point of the typical hydrocarbon mixture was defined, and we then predicted the solubility of hydrogen in such a component.

We have shown that the combination of MD simulations with the simpler EOS techniques can provide a rapid scheme for predicting hydrogen solubility in heavy oils accurately. The EOS method is fast and easy to use, with limited accuracy at high temperatures and pressures since EOS parameters are often state dependent. On the other hand, the MD simulations provide more accurate predictions at such conditions, but are difficult to implement and time consuming. A combination of these two techniques, as described here leads to a useful practical method.

(7b) Thermodynamic Study on Refrigeration System and Application Analysis

Jian Zhang and Qiang Xu, Dan F. Smith Department of Chemical Engineering, Lamar University, Beaumont, TX

A refrigeration system generally works by indirectly transferring the thermal energy from low-temperature sources to high-temperature sinks at the expense of electricity or mechanical work, such that the source temperature can be lowered or maintained at a certain value. In many industrial processes, the refrigeration system is employed for chilling and freezing the final liquid/solid products or liquefying the gas intermediates/products for the downstream cryogenic separation. It would be impossible for chemical/petrochemical process industries with the absence of the refrigeration system, such as the production of ethylene, propylene, oxygen, nitrogen, and ammonia. The performance of a refrigeration system is closely related to product quality and energy use efficiency, and thus the plant profitability. Meanwhile, because of the low-temperature operational characteristic, the refrigeration system inherently runs at a potential safety risk. Thus, in chemical/petrochemical industry practice, the performances of a refrigeration system under both normal and turnaround conditions are always the major concerns for the plant owners, engineers, and operators. Therefore, the operational optimality of a refrigeration system should be studied at the very beginning of the conceptual design stage. The design of a chemical plant refrigeration system is a very challenging task. The complexity mainly manifests in the following two aspects: i) An industrial refrigeration system usually needs to satisfy the cooling demands from many different units working at different temperatures. Thus, a refrigeration system should simultaneously work in different recycling loops at multiple temperature and pressure levels. Meanwhile, different refrigeration loops are highly interacted, and thus should be well integrated in terms of minimizing electricity or mechanical work consumptions. ii) A complex refrigeration system sometimes needs to accomplish different refrigeration tasks covering a wide temperature range. For instance, the refrigeration system at ethylene plants needs cooling different process streams changing from 23 to -100.8 Celsius degree. Thermodynamically, it is impossible by using only one refrigerant as the sole working media. Thus, multiple refrigerants have to be employed under this situation, such that each refrigerant will have its own working loops and cover the refrigeration tasks within a specific temperature range. Meanwhile, different recycling loops from different refrigerants may be optimally and cascadely integrated to jointly accomplish some refrigeration tasks. Note that there is still lacking of the systematical study on the optimal design of chemical industrial refrigeration system with the consideration of integrated multi-refrigerants and multi-working loops simultaneously. A typical example with such a complexity is the refrigeration system of an ethylene plant, where the refrigerants of ethylene and propylene are cascaded and coordinated to work in multiple loops for the cryogenic separation of many process streams. To address the conceptual design of such a cascade refrigeration system under various industrial constraints, a systematic optimization model containing multi-refrigerant and multi-temperature loops have to be developed and solved. In this paper, a general MINLP (mixed-integer linear programming) model for the optimal synthesis of the cascade refrigeration system in ethylene plants has been developed, where multiple refrigerants with multiple recycling loops are simultaneously addressed. Meanwhile, the exergy-temperature (B-T) chart combined with the exergy analysis, for the first time, is presented to comprehensively analyze the thermodynamic nature of a refrigeration system, which provides a solid foundation for the conceptual design of a complex refrigeration system. The efficacy of the developed methodology is demonstrated through a case study with industrial data.

See more of this Session: University Research In Energy Field for Industry
See more of this Group/Topical: Topical 5: University Research for Industry

SELECTIVE BROMINATION OF METHANE OVER SOLID ACID CATALYSTS AND POLY (4-VINYL PYRIDINE) CATALYZED HYDROLYSIS AND METHANOLYSIS OF METHYL BROMIDE

G. K. Surya Prakash, Patrice Batamack, Juan C. Colmenares, Thomas Mathew, George A. Olah
gprakash@usc.edu, batamack@usc.edu, jcarlos@ichf.edu.pl, tmathew@usc.edu, olah@usc.edu
Loker Hydrocarbon Research Institute and Department of Chemistry, University of Southern California, Los Angeles, CA 90089-1661, USA

Abstract:

The direct conversion of methane to methanol remains a big challenge for the scientific community. The large scale production of methanol is still achieved through the high temperature and capital intensive steam reforming of methane to synthesis gas followed by the reduction of CO under pressure. Several routes have been explored in the conversion of methane to valuable chemicals. Among these routes halogenation has been considered as an alternative way to transform methane to methanol/dimethyl ether via the hydrolysis of the methyl halide. Because bromination is slightly exothermic and more selective than chlorination and hydrogen bromide can be easily oxidized to bromine, we carried out bromination of methane with bromine over various solid acids under flow conditions. Methyl bromide was obtained in selectivity ranging from 70% to 100%. Methyl bromide was subsequently transformed to methanol/dimethyl ether using poly (4-vinyl pyridine) as a catalyst and an acid scavenger under moderate reaction conditions (100 °C, 8 bar).

Keywords: Bromination of methane, methyl bromide, methanol, dimethyl ether.

Introduction

Methane, the most abundant component of natural gas, can be a source of valuable compounds. But, its high stability generally requires harsh conditions for its functionalization.[1-6] The industrial production of the synthesis gas through the steam reforming of methane is one of the indirect routes to useful chemicals. Although leading to high yields, this route is highly capital and energy intensive. Research have been undertaken to find direct ways and mild conditions to functionalize methane. Direct routes are so far confronted to thermodynamic and kinetic limitations leading to low yields. Strong oxidants H_2O_2,[7,8] O_3,[7,] SO_3,[9] H_2SO_4,[10] Cl_2,[7,11-14] and Br_2[7,15-17] are often used to react with methane to achieve functionalization under mild conditions. Despite of its limited availability, we have focused our research using bromine for its moderate reactivity compared to other oxidants with the aim of achieving high selectivity to methyl bromide. Methyl bromide can be converted to methanol by hydrolysis, to ethers with alcohols or oligomerized to ethylene and higher hydrocarbons. The facile oxidation of hydrogen bromide, the by-product of these reactions, allows the recovery of bromine. We report here the results obtained in the bromination of methane in a flow system over various solid catalysts: Pt/SiO_2, Pt/Al_2O_3, Rh/Fumed SiO_2, 5% Rh/11% Fe/Fumed SiO_2, H-Beta, USY, H-ZSM-5, and Silicalite. Subsequently, methyl bromide was reacted with water or methanol to yield methanol and dimethyl ether under batch conditions using poly (4-vinyl pyridine) as a catalyst and an acid scavenger.

Experimental

Materials:

Ultra high purity (99.99%) methane was obtained from Airgas. Bromine (purity > 99.5%) was purchased from Sigma-Aldrich and 5% Pt/Al_2O_3 from Alfa Aesar. The powder form of HZSM-5 (Si/Al = 140) and USY (Si/Al = 5.2) were generously donated by Zeolyst International. H-Beta and Silicalite zeolite samples were generously provided by UOP (Honeywell) in powder forms.

The metal supported catalysts were prepared by impregnation. A mixture of water and methanol was used as the medium. Pt/SiO_2 was prepared by impregnation from a solution of H_2PtCl_6 on silica gel (Davidson) and calcined in air at 400 °C for 5 h. The 2% Rh/fumed SiO_2 catalyst was prepared by impregnation of a solution of $RhBr_3$ on fumed silica with appropriate amounts. Similarly, 5% Rh/11% Fe/fumed SiO_2 was prepared by successive impregnations of alcoholic solutions of $FeBr_2$ and $RhBr_3$ on fumed SiO_2 (Sigma-Aldrich) followed by calcination in air at 350 °C for 2 h. The catalysts were reduced under H_2 flow for 1 h prior to the reaction.

For hydrolysis and methanolysis, all reagents were purchased from Sigma-Aldrich and used without further purification. The purity of methyl bromide was 99.5%. Anhydrous methanol purity was 99.9% (water < 0.005%). PVP, 2% cross-linked with divinyl benzene having a surface area of 17 m^2/g was used for the reactions.

Catalytic test reactions:

Bromination experiments were performed on a 620 mm long quartz tubular reactor with a diameter of 9 mm. In a typical experiment with the long quartz tube, catalyst (3 g) was placed in the reactor supported by quartz wool. The catalyst was heated in air from room temperature to 500 °C at a rate of 3 °C/min and kept for 1 h at that temperature in the case of zeolites. Then, the catalyst was cooled down to the reaction temperature under nitrogen or helium flow. The reactants were fed from the top downward when the catalyst load was less than 5 g and bottom upward for 5 g of catalyst and more to minimize the pressure buildup in the reactor. The methane flow, controlled by a mass flow controller (from Aalborg), entrained bromine vapor from a bromine reservoir maintained at either 2 °C or 25 °C to achieve a given methane to bromine ratio of 2.94 to 1 or 10.3 to 1. After 3 h on-stream, the products from the exit of the reactor were collected for 1 h in a solution of $CDCl_3$ kept at -60 °C and analyzed by proton NMR using dichloromethane as an internal reference. The connection from the exit of the reactor to the sample collector was heated at 150°C to avoid condensation of the products.

Hydrolysis and methanolysis experiments were carried out in a closed pressure tube (internal volume, 20 mL) between 75 °C and 125 °C for 2 h. Further details are provided in references 18 and 19.

Results and discussion

1. Bromination

1.1. Supported noble metal catalysts:

Table 1 shows that thermal bromination generally proceeds with the formation of three bromomethane derivatives (CH_3Br, CH_2Br_2, $CHBr_3$). The selectivity to methyl bromide was improved with the use of catalysts. Among the supported noble metals, 2% Rh on fumed silica exhibited the highest activity and a good selectivity to methyl bromide. The addition of iron was detrimental. In all cases, the leaching of the metal was observed at the bottom of the reactor and in the case of rhodium on alumina, sublimation of the support occurred. Carbon deposits were visible on all catalysts.

Table 1: Bromination of methane over noble metal on silica and alumina

Catalyst	Blank	Rh (2%) on fumed silica	Rh (5%) Fe (11%) on fumed silica	Pt (4.5%) on silica	Pt (5%) on alumina
Temp. (°C)	435	435	435	415	415
Br_2 Conversion (mol %)	52.5	59.8	38.1	32.3	34.7
		Yield (mol %)			
CH_3Br	26.8	53.2	28.1	23.9	30.5
CH_2Br_2	23.9	6.6	10.0	0.8	4.2
$CHBr_3$	1.8	-	-	-	-

Reaction conditions: CH_4/Br_2 molar ratio 2.94:1; 3g of catalyst; GHSV ($mL.g^{-1}.h^{-1}$) = 100.

1.2. Zeolites

HZMS-5 zeolites are known to be stable catalysts for the oxychlorination of methane. As expected, Table 2 shows that the selectivity to methyl bromide is improved in the thermal bromination at high methane to bromine ratio (10 to 1). However, the 3 bromomethanes derivatives can still be observed. Under these conditions, methyl bromide was the sole product observed when zeolites are used as catalysts. H-Beta was the most active among the tested zeolites. At lower methane to bromine ratios, the conversion of bromine (and therefore that of methane) was high with a high selectivity to methyl bromide. But, the 5 h test with HZSM-5 and the 24 h test with Silicalite showed a loss of activity of the catalyst with time due to the dealumination of the framework by the action of HBr as previously observed in the chlorination of methane with zeolites wherein HCl is produced.[12]

Table 2: Bromination of methane over various zeolites.

Catalyst	Blank	H-Beta	USY	HZSM-5	HZSM-5	Silicalite
Weight (g)	-	1	3	5	5	5
CH_4/Br_2 (mol ratio)	10	10	10	10	1.2[1]	3
GHSV ($mL.g^{-1}.h^{-1}$)	7.2[2]	438	144	84	84	84
Temp. (°C)	370	370	370	370	435	435
Br_2 Conversion (mol %)	57.8	61.0	46.4	63.0	50.4 (49.1)[3]	78.5 (74.7)[4]
		Yield (mol %)				
CH_3Br	46.8	61.0	46.4	63.0	42.2 (34.8)[3]	73.0 (61.6)[4]
CH_2Br_2	7.3	-	-	-	8.3 (13.6)[3]	5.2 (13.1)[4]
$CHBr_3$	3.7	-	-	-	- (0.7)[3]	-

Reaction conditions: [1] Br_2 fed using a syringe pump; [2] CH_4 Flow rate ($mL.min^{-1}$); [3] 5 h on-stream; [4] 24 h on-stream.

2. Hydrolysis and Methanolysis of Methyl bromide to Methanol and Dimethyl ether using poly (4-vinyl pyridine) (PVP)

Methyl bromide can be converted to methanol and dimethyl ether by hydrolysis or methanolysis. The reaction is not thermodynamically favorable. But we found that under batch conditions PVP can act as a very efficient catalyst as well as an acid scavenger for selective hydrolysis and selective methanolysis of methyl bromide to respectively methanol and dimethyl ether (Tables 3, 4). As an acid scavenger, PVP helps drive the reaction towards the formation of cleaner products with decreasing hazardous wastes and pollution. While methanolysis of methyl bromide produced only dimethyl ether, hydrolysis yielded both methanol and dimethyl ether, which can be formed through reactions (2) and (3) in scheme 1.

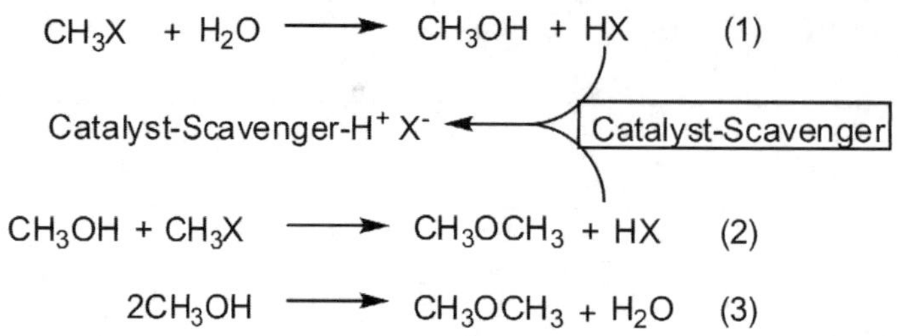

Scheme 1. Hydrolysis of methyl bromide to methanol and related reactions in the presence of PVP.

Table 3. Hydrolysis of methyl bromide using poly(4-vinyl pyridine)

Molar Ratio (CH_3Br:PVP)	Conversion of CH_3Br (%)	Selectivity (%) DME	MeOH
1:0	29.9	48.4	51.6
1:0.125	67.2	59.8	40.2
1:0.2	87.6	43.0	57.0
1:0.33	86.8	30.0	70.0
1:1	85.6	29.0	71.0

Reaction conditions: 2 h at 100°C with H_2O:CH_3Br molar ratio of 10:1.

Table 4. Methanolysis of methyl bromide using poly (4-vinyl pyridine)

Molar ratio (CH_3OH:CH_3Br:PVP)	Temp (°C)	Conversion of CH_3Br (%)	Selectivity of DME (%)
3:3:1	75	0.0	0.0
3:3:1	100	23.8	100.0
3:3:1	125	35.3	100.0
3:3:0	125	0.0	0.0

Reaction time 2h.

Conclusion

Selective monobromination was achieved over supported noble metal and zeolites in a flow system between 370 °C and 435 °C. At a methane to bromine ratio of 10, zeolites were able to produce methyl bromide as the sole product. At low methane to bromine ratios high conversion of methane were obtained with a good selectivity to methyl bromide. However, all catalysts suffered from deactivation after 3 to 5 h on-stream due to metal leaching by the action of HBr. Poly (4-vinyl pyridine) was shown as an efficient catalyst in the conversion of methyl bromide to methanol and dimethyl ether in the presence of water or methanol.

Acknowledgements

We are thankful to Department of Energy for the financial support and UOP (Honeywell) for providing catalyst material and to Zeolyst International for providing the H-ZSM-5 samples.

References

1. Gesser HD, Hunter NR. A review of C-1 conversion chemistry. Catal. Today 1998;42:183-189.
2. Lunsford JH. Catalytic conversion of methane to higher hydrocarbons. Catal. Today 1990;6:235-259.
3. Lunsford JH. The catalytic oxidative coupling of methane. Angw. Chem. Int. Ed. Engl. 1995;34:970-980.
4. Crabtree RH. Aspects of methane chemistry. Chem. Rev. 1995;95:987-1007.
5. York APE, Xiao T, Gree LH. Brief overview of the partial oxidation of methane to synthesis gas. Topics in Catal. 2003;22:345-358.
6. Parkyns ND, Warburton CL, Wilson JD. Natural gas conversion to liquid fuels and chemicals: Where does it stand? Catal. Today 1993;18:385-442.
7. Olah GA. Electrophilic Methane Conversion. Acc. Chem. Res. 1987;20:422-428.
8. Suss-Fink G, Nizova GV, Stanislas S, Shul'pin GB. Oxidations by the reagent'O_2-H_2O_2-vanadate anion-pyrazine-2-carboxylic acid'. Part 10. Oxygenation of methane in acetronitrile and water. J. Mol. Catal. A: Chemical 1998,130:163-170.
9. Mukhopadhyay S, Zerella M, Bell AT. A high-yield, liquid-phase approach for the partial oxidation of methane to methanol using SO_3 as the oxidant. Adv. Synth. Catal. 2005;347:1203 –1206.

10. Periana, R. A.; Taube, D. J.; Gamble, S; Taube, H.; Satoh, T.; Fujii, H. Platinum catalysts for the high-yield oxidation of methane to a methanol derivative. Science 1998,280:560-564.

11. Olah GA, Mo YK. Electrophilic Reaction at Single Bonds. X1II.Chlorination and Chlorolysis of Alkanes in SbF_5-$C1_2$-SO_2ClF Solution at Low Temperature. J. Am. Chem. Soc. 1972,94:6864-6865.

12. Bucsi I, Olah GA. Selective monochlorination of methane over solid acid and zeolite catalysts. Catal. Lett. 1992,16:27-38. (h) Olah GA, Gupta B, Farina M, Felberg JD, Ip WM, Husain A, Karpeles R, Lammertsma K, Melhotra AK, Trivedi NJ. J. Am. Chem. Soc. 1985,107:7097-7105.

13. Batamack P, Bucsi I, Molnar A, Olah GA. Electrophilic chlorination of methane over superacidic sulfated zirconia. Catal. Lett. 1994,25:11-19.

14. van Iersel MM, van Schilt MA, Benes NE, Keurentjes JTF. Controlled methyl chloride synthesis at mild conditions using ultrasound irradiation. Ultasonics Sonochemistry 2010,17:315-317.

15. Lorkovic IM, Yilmaz A, Yilmaz G A, Zhou X-P, Laverman LE, Sun S, Schaefer DJ, Weiss M, Noy ML, Cutler CI, Sherman JH, McFarland EW, Stucky GD, Ford PC. Catal. Today 2004;98:317-322.

16. Degirmenci V, Uner D, Yilmaz A. Methane to higher hydrocarbons via halogenation. Catal. Today 2005,106:252–255.

17. Degirmenci V, Yilmaz A, Uner D. Selective methane bromination over sulfated zirconia in SBA-15 catalysts. Catal. Today 2009,142:30–33.

18. Prakash G KS, Colmenares JC, Batamack PT, Mathew T, Olah GA. Poly(4-vinylpyridine) catalyzed hydrolysis of methyl bromide to methanol and dimethyl ether. J. Mol. Catal. A:Chem. 2009,310:180-183.

19. Prakash G KS, Colmenares JC, Batamack PT, Mathew T, Olah GA. Poly(4-vinylpyridine) catalyzed selective methanolysis of methyl and methylene bromides. Tetrahedron Lett. 2009,50: 6016-6018.

(7d) Exploiting Metal-Support Interactions to Optimize Dispersion and Reducibility of a Highly Active & Selective Fischer-Tropsch Synthesis Eggshell Catalyst

Lucky R. Landrigan[1], Syed Ali Zeeshan Gardezi[1], Babu Joseph[2] and John Wolan[2], (1)Chemical Engineering, University of South Florida, Tampa, FL, (2)Chemical & Biomedical Engineering, University of South Florida, Tampa, FL

In this study the optimum solution chemistry for synthesis of a highly active and selective Fischer-Tropsch (FT) eggshell cobalt-based catalyst is investigated. The precursor salt ($Co(NO_3)_2$*$6H_2O$) is dehydrated and dissolved in anhydrous ethanol under a low humidity nitrogen atmosphere and titrated with anhydrous ethanol-urea solution. It was found that the addition of small amounts of de-ionized water followed by titration ionizes the solution and greatly enhances active metal dispersion on the silica gel support. However, higher amounts of added water can either aid in the ionization of salt or lead to the recombination of ions during titration resulting in higher or lower metal-to-surface interaction respectively.

The silica gel surface can become negatively charged due to ion exchange with an ethanol-water mixture. If the water is present in small quantities, it will not form an immobilized aqua monolayer, allowing for vertical stacking of ethanol on the gel to leave vacant aligned spaces for salt attachment. After titration with urea, when a small amount of water is present, there is a net increment of salt ions which interact with the silica gel to form coordinated covalent bonds and eventually lead to higher dispersions. On the other hand, if the quantity of water is high, an immobilized aqua monolayer is formed on the silica gel before ethanol can interact. Also, this large amount of water leads to recombination of ions, thus reducing the free floating Co^{2+} ions near the surface. Under such circumstances, direct metal support interaction is not possible leading to poor non uniform metal distribution on the surface and eventual sintering.

In order to exploit this interaction, the optimum urea-water-ethanol proportions were found via conductivity tests of solutions throughout the titration process. Hydrogen chemisorption is used to test this hypothesis.

See more of this Session: University Research In Energy Field for Industry
See more of this Group/Topical: Topical 5: University Research for Industry

(7e) Effect of Scale-up on the Heat Transfer Coefficient and Rate In Bubble Columns

Rahman S. Abdulmohsin[1], Balasim A. Abid[2] and Muthanna H. Al-Dahhan[1], (1)Department of Chemical and Biological Engineering, Missouri University of Science and Technology, Rolla, MO, (2) Department of Chemical Engineering, University of Technology-Baghdad, Baghdad, Iraq

Scale-up effects on the time-averaged local heat transfer coefficients were investigated in two different size columns with diameters of 0.16 and 0.44 m. Air-water system was used with superficial gas velocities ranging from 0.03 to 0.30 m/s which cover both the homogenous and heterogeneous flow regimes. By using an advanced fast response heat transfer probe, the local instantaneous heat transfer coefficients were measured in fully developed flow region (Z/D=5.5) at different radial positions of the columns. These measurements are important to select proper location and appropriate configuration of internal heat exchanging surfaces in bubble columns. Experiments results indicate that time-averaged local heat transfer coefficients in 0.44 m column diameter are larger than these obtained in 0.16 m column diameter. The heat transfer coefficients in the center of both columns for all the conditions studied are larger than those near the wall region. Also, local heat transfer coefficient increases as superficial gas velocity increased.

Keywords: *bubble column; heat transfer coefficients; column diameter effect; scale-up.*

Note: the work was performed at Washington University in St. Louis.

See more of this Session: University Research In Energy Field for Industry
See more of this Group/Topical: Topical 5: University Research for Industry

(23a) Dramatic Nano-Fluidic Properties of Carbon Nanotube Membranes and Application In Chemical Processes and Water Treatement

Bruce J. Hinds, Chemical and Materials Eng., University of Kentucky, Lexington KY, KY

Carbon nanotubes have three key attributes that make them of great interest for novel membrane applications 1) atomically flat graphite surface allows for ideal fluid slip boundary conditions 2) the plasma cutting process to open CNTs inherently places functional chemistry at CNT core entrance and 3) CNT are electrically conductive allowing for electrochemical reactions and application of electric fields gradients at CNT tips. To explore the hypothesis of 'Gatekeeper' membrane selectivity, the entrances to CNT's cores were functionalized with aliphatic amines of different lengths, charged dye molecule and poly-peptides and a hindered diffusion model with a geometry of CNT tip functionalization was consistent with the experimentally observed selectivities. Changes in the flux and selectivity also support a model where charged tethered molecules at the tips are drawn into the CNT core at positive bias hindering/gating flux across the membrane. In general, the transport mechanisms through CNT membrane are a) ionic diffusion is near bulk expectation with no enhancement from CNT b) gas flow is enhanced by ~1-2 order of magnitude due to specular reflection off of flat graphitic surface c) and pressure driven flux of a variety of solvents are 4-5 ORDERS OF MAGNITUDE FASTER than conventional Newtonian flow due to atomically flat graphite planes inducing nearly ideal slip conditions. This is a nearly ideal platform to induce electro-osmotic flow with high charge density at pore entrance and a nearly frictionless surface for the propagation of plug flow. Through diazonium electrochemical modification we have successfully bound anionic surface charge to CNT tips and along CNT cores. High electro-osmotic flows of 0.05 cm/s at -300mV bias are seen with 40 fold improvements in power efficiency. Initial application demonstration in programmed drug delivery, catalytic conversion platform, bioseparations and water treatment are highlighted.

See more of this Session: University Research In Water, Bioseparation, and Pharmaceutical Field for Industry
See more of this Group/Topical: Topical 5: University Research for Industry

(23b) Fouling Resistant Membranes

Benny D. Freeman, Chemical Engineering, University of Texas at Austin, Austin, TX

An urgent need in the 21st century is access to secure, sustainable sources of fresh water. Membrane-based desalination has been recognized as an energy efficient, low environmental impact route to produce clean water for a variety of applications. Additionally, ultrafiltration and microfiltration of aqueous streams have become important unit operations in many industries, including the water purification and wastewater purification industries. This presentation focuses on 2 new classes of polymers to attack critical problems in membrane-based water purification: (1) chlorine-tolerant desalination membranes and (2) fouling-resistant membrane nanolayers based on polydopamine.

Commercially available RO membranes are prepared mainly from aromatic polyamides (PA). However, PA membranes suffer from poor resistance to continual exposure to oxidizing agents such as chlorine. Chlorine is the most widely used oxidizing biocide in water treatment because it is inexpensive and highly effective when present in water at levels of a few ppm. Disinfection of feed water to membrane desalination units is required to prevent biofilm growth on the membranes, which significantly degrades their performance. However, PA membranes cannot tolerate continuous exposure to water containing more than a few ppb of chlorine. This presentation discusses results from a systematic study of the desalination properties of a new family of highly chlorine tolerant, sulfonated polysulfones. These polymers are prepared via a novel synthesis involving direct copolymerization of non-sulfonated monomers with a disulfonated monomer. By controlling the content of disulfonated monomer in the final polymer, desalination properties, such as water permeability and salt rejection, can be varied over a wide range. A family of materials, based on random copolymers, phase-separated block copolymers, blends, and crosslinked materials may be prepared from these starting blocks to provide a new platform for desalination membranes. We are currently working with a major producer of desalination membranes to reduce these new materials to practice.

Across many platforms of membranes, fouling mitigation is a major challenge to be addressed to achieve the most energy-efficient, cost-effective membrane filtration processes. Previously, many surface modifications and functionalized polymers were reported to prevent fouling. However, most of these techniques and materials are practically difficult to implement in water purification membranes. We have discovered surface treatment methodologies that can be used to prepare high permeability polymeric membranes from all common water purification membrane classes. These surface-modified membranes have persistent tolerance to fouling by proteins and emulsified oil, two ubiquitous contaminants in a variety of wastewaters. These membranes were prepared by depositing bio-inspired, self-polymerized hydrophilic polydopamine nanolayers on their surfaces. To demonstrate scalability of this modification, the nanolayers were also applied to entire membrane modules. Upon nanolayer deposition, the membranes could be further functionalized using fouling-resistant macromolecules, such as poly(ethylene glycol), to promote improved fouling resistance and, therefore, high membrane flux. The simplicity, versatility, and broad applicability of the nanolayer deposition technique on membranes provide distinct advantages over other modification strategies to reduce fouling. A start-up company, Advanced Hydro, Inc., based in Austin, Texas has been formed to commercialize these approaches to improve membrane fouling resistance.

See more of this Session: University Research In Water, Bioseparation, and Pharmaceutical Field for Industry
See more of this Group/Topical: Topical 5: University Research for Industry

(23c) Synthesis of Reactive Nanoparticles In Membranes

D. Bhattacharyya[1], V. Smuleac[1] and Scott R. Lewis[2], (1)Chemical and Materials Engineering, University of Kentucky, Lexington, KY, (2)Chemical & Materials Engineering, University of Kentucky, Lexington, KY

Nanostructured metals have become an important class of materials in the field of catalysis, optical, electronic, magnetic and biological devices due to the unique physical and chemical properties. The nanosized metals precipitated from solutions are extremely reactive due to the high surface energy, and they usually form aggregates without the protection of their surface. Therefore, immobilization of metal nanoparticles in polymer membrane (such as, cellulose acetate, PVDF, polysulfone, chitosan, etc,) media is important from the point of view of reactivity, organic partitioning, preventing loss of nanoparticles, and reduction of surface passivation. Another major advantage of having a polymer domain is that nanoparticles (without causing agglomeration) can be directly synthesized in the matrix. This presentation will include technology developments relating to direct synthesis (including "green chemistry" approaches) of iron and bimetallic nanoparticles in microfiltration membranes for environmental applications. The project has been funded by NIEHS , US. EPA, and DOE-KRCEE.

See more of this Session: University Research In Water, Bioseparation, and Pharmaceutical Field for Industry
See more of this Group/Topical: Topical 5: University Research for Industry

(23d) The Design of Advanced Functional Membranes for Industrial Bioseparations

Scott M. Husson, Department of Chemical and Biomolecular Engineering, Clemson University, Clemson, SC

Market demand is increasing rapidly for biotherapeutics such as recombinant proteins, monoclonal antibodies, viral vaccines and plasmid DNA. With the advent of molecular biotechnology and engineered cell lines, upstream production processes have made unprecedented progress in the last decade, shifting the production burden to the downstream processing. To improve process economics and meet market demand, manufacturers will require higher productivity and higher resolution separation techniques – membrane chromatography fulfills these requirements.

In this talk, I will describe our efforts to develop adsorptive membranes for use as chromatography media. I will describe the design and performance characterization of these materials and dispel two common misperceptions by showing that (1) membrane chromatography can be a higher capacity process than resin chromatography in the purification of biologics, and (2) membrane chromatography can be a higher resolution process than resin chromatography.

Our design approach is to graft polyelectrolyte nanolayers from the pore surfaces of macroporous membranes using a controlled polymerization method. Using a number of examples, I will demonstrate that the dynamic binding capacities of proteins and DNA can be dramatically higher for well-designed macroporous membrane adsorbers than for existing commercial membrane adsorbers and resin columns. Using controlled polymerization and high polymer chain density produces adsorptive membranes with high binding capacities that are independent of flow rate, enabling high-throughput operation.

See more of this Session: University Research In Water, Bioseparation, and Pharmaceutical Field for Industry
See more of this Group/Topical: Topical 5: University Research for Industry

(23e) Microfluidic Platforms to Screen for Crystallization Conditions of Active Pharmaceutical Ingredients

Sachit Goyal[1], Michael R. Thorson[2], Yuchuan Gong[3], Geoff G.Z. Zhang[3], Charles F. Zukoski[4] and Paul J. A. Kenis[5], (1)Department of Chemical & Biomolecular Engineering, University of Illinois at Urbana-Champaign, Urbana, IL, (2)Department of Chemical & Biomolecular Engineering, University of Illinois at Urbana Champaign, Urbana, IL, (3)Solid State Sciences, Abbott Laboratories, North Chicago, IL, (4)Department of Chemical and Biomolecular Engineering, University of Illinois at Urbana Champaign, Urbana, IL, (5)Chemical & Biomolecular Engineering, University of Illinois at Urbana-Champaign, Urbana, IL

Screening for polymorphs is critical in the API development process. Identifying the available solid forms of candidate drugs (CD) as early as possible require intensive screens varying multiple parameters including counter ions, antisolvent and concentration of the CD.

Presently, highly complex robotic systems are used at the macroscale that utilize sample volumes of the order of 5 µl per condition [1,2]. Due to small quantity of the CD available in the early stages, number of conditions that can be screened is limited. Microfluidics have the potential to improve high throughput screening by utilizing reduced sample volumes, allowing improved control over mixing of fluids via free interface diffusion, and increasing the number of conditions that can be screened simultaneously [3].

Here we present a microfluidic platform enabling high throughput screening (48 wells on one platform) of antisolvents for polymorphism and crystal habit of APIs utilizing extremely small sample volumes (50nl per well). Several parameters like ratio of volume of API to antisolvent (varied from 1:5 to 5:1), solvent, and API concentration are varied to control supersaturation and are found to influence crystal habit and polymorphism. The microfluidic platform has been modified to enable in-situ Raman spectroscopy for the analysis of crystals and validation of polymorphs crystallized on-chip without need to manually harvest crystals from the chip. To support experimental results, we modeled the effect of geometry of wells on the supersaturation profiles and hence on the nucleation kinetics of the API with antisolvent. We demonstrate the capability of the chip using two model APIs – Indomethacin and Naproxen to generate different polymorphs and crystal morphologies on chip.

References

[1] P. Desrosiers, *Modern Drug Discovery*, 2004, 40.

[2] S. Talreja, D.Y. Kim, A.Y. Mirarefi, C.F. Zukoski, P.J.A. Kenis, J. Appl. Crystallography, 2005, 38, 988-995.

[3] C. Hansen, E. Skordalakes, J. Berger, S. Quake, *Proc. Nat'l Acad. Sci.*, 99, 2002, 16531-6.

See more of this Session: University Research In Water, Bioseparation, and Pharmaceutical Field for Industry
See more of this Group/Topical: Topical 5: University Research for Industry

An efficient p-chlorophenol removal prepared from cationic gemini surfactant modified montmorillonite

Xuemin Liu[1,2], Chunxu Lu[2], Cong Song[1], Songying Wang[1], Yi Wang[1]

(1.School of Chemical and Material Engineering, Jiangnan University, Wuxi Jiangsu, 214122, China; 2.School of Chemical Engineering, Nanjing University of Science and Technology, Nanjing Jiangsu, 210094, China)

Corresponding author. E-mail address: liu65xm@yahoo.com.cn

Abstract: A group of Gemini cationic surfactants containing three hydrophobic and two hydrophilic groups were used as the spacer to modify sodium montmorillonite (MMT-Na) via cation exchange to produce a *p*-chlorophenol removal. The Gemini cationic MMT was characterized with FT-IR and XRD, and its thermal stability was analyzed using thermogravimetric analysis (TG).

The modified montmorillonites performed a good removal of *p*-chlorophenol in wastewater. The *p*-chlorophenol adsorption amount of the Gemini cationic MMT is up to 39.36 mg/g, which is 7 times higher than that of MMT-Na. The modified montmorillonites also present a good thermal stability and the decomposition happened at 200-500°C mainly due to initial weight loss from residual water desorption, followed by decomposition of the surfactant and the dehydroxylation of structural water in the montmorillonite layers. The structure-function relationship of the surfactant was also investigated.

Keywords: Gemini cationic surfactant; Montmorillonite; modification; *p*-chlorophenol; adsorption

1. Introduction

Chlorophenols are widely used as pesticides and antiseptics. In addition to commercial production, chlorophenols can be produced when waste or drinking water is disinfected with chlorine. Various chlorophenols with isomers are commonly found organic pollutants in subsurface environments. Direct exposure to chlorophenols can be fatal. Long-term exposure to the chemicals may cause cancers and affect the function of the liver and the immune system. Numerous researchers have investigated methods to remove organic pollutants from aqueous phases (Lawrence et al., 1998; Wibulswas et al., 1999; Juang et al., 2004; Pernyeszi et al., 2006; Zampori et al., 2009). Adsorption of the pollutants using organo-montmorillonite has attracted much attention as one of the removal methods.

Cationic surfactants intercalated montmorillonite have gradually been the subject of extensive research, and conventional cationic surfactants were widely used as modifiers, while Gemini surfactants which contain at least two hydrophilic and two hydrophobic groups were seldom used. In fact, the montmorillonite modified by Gemini surfactants was proved to be more efficiency in removing the organic pollutes (Rosen and Li, 2001; Li and Rosen, 2000) than the one modified by conventional cationic surfactant.

In this paper, montmorillonites modified by Gemini cationic surfactants which contain three hydrophobic and two hydrophilic groups, were prepared and characterized by FTIR, XRD and TG analysis. The *p*-chlorophenol adsorption test was also carried out, expecting to be useful in removing organic pollution in wastewater.

2. Materials and methods

2.1 Materials

The montmorillonite used in this study was from ZheJiang Huate Chemical Co. Ltd (Zhejiang, China). The cation exchange capacity (CEC) was 70 mmol/100 g. A new type of Gemini cationic surfactants in Fig.1, which were synthesized according to the literature (Chen et al., 2008a; Joanna and Jan., 2006), were used as modifiers.

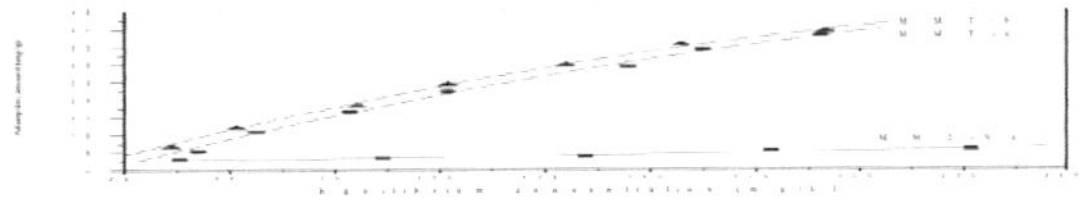

R1	n-$C_{11}H_{23}$	n-$C_{13}H_{27}$
R2	n-$C_{12}H_{25}$	n-$C_{14}H_{29}$
Surfactants	a	b

Fig.1. The structure of Gemini cationic surfactants

2.2 Preparation of modified montmorillonites

The modified montmorillonites were prepared as follows: Sodium montmorillonite power (5g) was dispersed in 100 mL of deionized water. After adjusting pH to 7.4 with hydrochloric acid, the dispersion was stirred at 80°C for 1 h. Gemini surfactants (2mmol) were dissolved in 50mL of deionized water and slowly dropped into the montmorillonite dispersion. After dropping, the dispersion was treated at 80°C for 4 h. The precipitates were separated by filtration and washed several times with deionized water until no chloride ion was detected by $AgNO_3$ solution. In this paper, MMT-a, b represent montmorillonite modified by surfactant a, b, respectively.

2.3 Characterization methods

2.3.1 Fourier transform infrared spectroscopy

The Fourier transform infrared (FTIR) spectra were obtained on FLA2000-104 FTIR spectrometer.

2.3.2 Thermogravimetric analysis

Thermogravimetric analysis (TG) were performed on a TGA Instruments (TGA /SDTA 85/e) and operated at a rate of 10.00°C/min from room temperature to 800.00°C in a high-purity flowing nitrogen atmosphere (100.0 mL/min).

2.3.3 X-ray diffraction

X-ray diffraction measurements (XRD) were performed using a D8-Advance X-ray diffractometer with CuKα radiation (λ = 1.5418 Å), operated at 40 KV and 30mA in a step scan mode. Diffraction measurements were conducted within the range of 2–10°, at a scanning speed rate of 0.02°/s.

2.4 Adsorption test

The adsorption experiments were carried out using batch equilibrium tests. 0.1g modified montmorillonite and 25 mL of prepared solution containing *p*-chlorophenol were placed in 250 mL conical flask. The adsorption experiments were carried out at

(35±0.2) °C and stirred at 180 r/min for 60 min. After separated by centrifugation, the concentration of *p*-chlorophenol was determined by Double Beam UV-VIS Spectrophotometers (λ_{max}= 280 nm).

3. Results and discussion

3.1 Characterization of modified montmorillonites

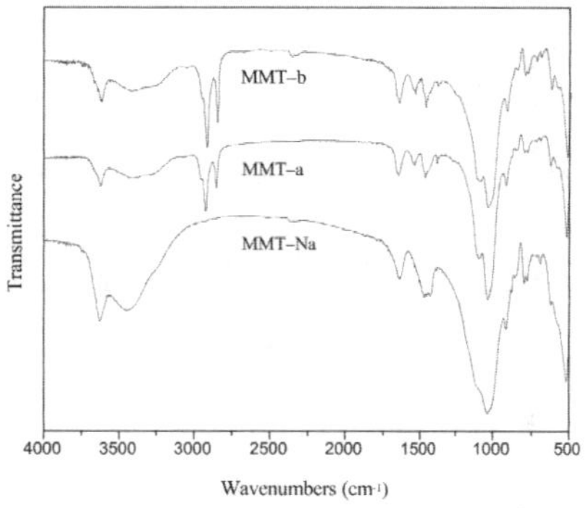

Fig.2. IR spectra of MMTs

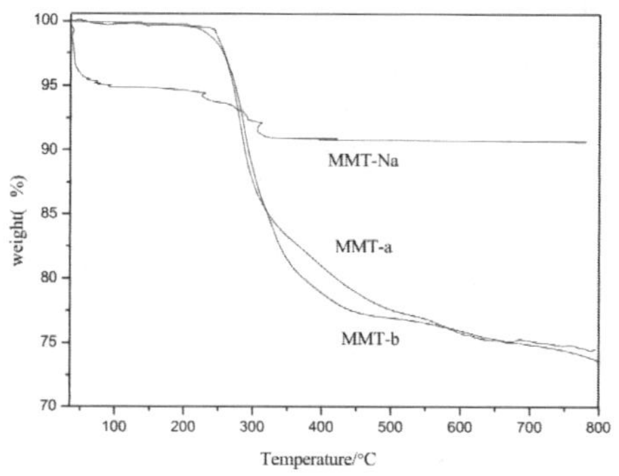

Fig.3. TG curve of MMTs

The unmodified montmorillonite is characterized by the bands of Si-O stretching and bending as well as OH bending (Zhou et al., 2009). The C-H stretching vibration of alkyl chains usually occur in the 3020-2800 cm^{-1} region (Kozak and Domka., 2004). In Fig.2, OH stretching vibration and bending vibration appears at 3640~3440 cm^{-1} region and 1639 cm^{-1} respectively. The absorption band at 1030 cm^{-1} is attributed to Si-O stretching vibration, while bands in 600~500 cm^{-1} are assigned to Si-O and Al-O bending vibration. The peaks at 2923 $cm^{-1}$and 2850 cm^{-1} are due to the asymmetrical and symmetrical C-H stretching of the methylene groups. The C-H bending vibration of the alkyl chains is observed at 1468 cm^{-1}. As is the Characteristic absorption peak of amide group, the absorption peaks that occur in 1652 cm^{-1} and 1545 cm^{-1} must be emphasized. Since there is no chloride ion to be detected, it can be proved that the surfactants have been successfully intercalated into the interlayer of montmorillonite.

Table 1. The weight loss in the thermogravimetric curves of MMTs

Samples	We.% loss at 35-200°C	We.% loss at 200-500°C	We.% loss at 750°C
MMT-Na	5.43	3.80	9.31
MMT-a	0.26	22.14	25.23
MMT-b	0.38	22.69	25.57

Thermo-analytical method provides a measure of the thermal stability of materials (Chen et al., 2008b) and has been widely used in studying the thermal decomposition of the modified montmorillonite(Xi et al., 2005a,b; Xie et al., 2001, 2002; He et al., 2005; Zhu et al., 2008). As shown by the TG curve in Fig.3, the thermal decomposition of MMT-Na occurs in two steps. The first step of weight loss is observed at 25-200°C region which is attributed to desorption of free water (water between particles and sorbed on the external surfaces of crystals) from the montmorillonite. The second step of weight loss occurs in 250-450°C regions, corresponding to the loss of water in the interlayer or to the dehydration of the hydrated cation in the interlayer. In Table 1, the organic constituent in modified montmorillonite mainly decompose at 200-500°C region, while there is little weight loss at 35-200°C region, which indicates that the free water has been replaced by Gemini cationic surfactants.

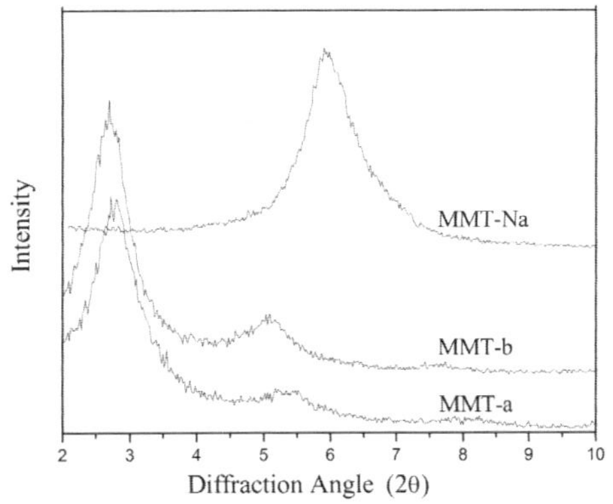

Fig.4. X-ray diffraction patterns of MMTs

According to Fig.4, the layer distance (Δd) was calculated by subtracting the sickness of the silicate layer of montmorillonite which equals 0.96 nm (Huskic et al., 2008), and listed in Table 2.

Table 2. The maximum diffraction angle and basal spacing of MMTs

Sample	Maximum diffraction angle 2θ/(°)	Basal spacing d001 (nm)	Layer distance Δd(nm)
MMT-Na	5.94	1.47	0.51
MMT-a	2.80	3.15	2.19
MMT-b	2.68	3.29	2.33

From Table 2, the basal spacing value of MMT-a and MMT-b is 3.15 nm and 3.29 nm respectively. As the carbon chain length of Gemini cationic surfactants increase, the size of the modifier augment, and the basal spacing also increases. Compared with the modifier Gemini cationic surfactant (12-4-12) reported by Zhou et al (2009), which increases the basal spacing to 2.06 nm, the modifiers used in this paper are more effective at expanding the interlayer space of the montmorillonite. According to literature (Lagaly, 1986), the alkyl chains in the modified montmorillonite interlayer might form monolayers(1.37nm), bilayers(1.77nm), pseudotrimolecular layers(2.17nm) and paraffin structure(2.77nm) as the basal spacing changes. We can conclude that these novel Gemini surfactants with more hydrophobic space groups are easier to form a paraffin structure in the montmorillonite interlayer than 12-4-12.

3.2 The adsorption isotherms of *p*-chlorophenol on montmorillonites

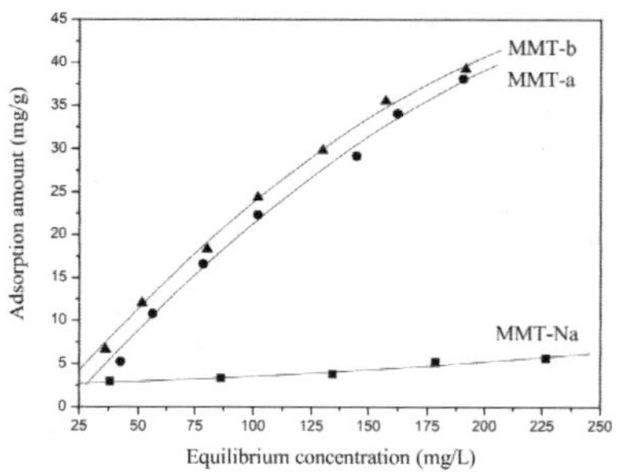

Fig.5. Isotherms of the adsorption of *p*-chlorophenol on MMTs

As shown in Fig 5, the adsorption amount of *p*-chlorophenol on MMT-Na has a linear relationship with equilibrium concentration, and the growth rate is extremely low. The modified montmorillonite did not have advantage in adsorption to *p*-chlorophenol at low equilibrium concentration, however, as the equilibrium concentration increase, the adsorption amount to *p*-chlorophenol of the modified montmorillonite increases rapidly. The adsorption amounts are up to 39.36 mg/g and 39.27 mg/g respectively at the maximum equilibrium concentration, which is seven times higher than the maximum adsorption amount of MMT-Na. It can be explained that there be more adsorption area and surface non-polarity of the montmorillonites modified by these novel Gemini surfactants. For the surface of MMT-Na is hydrophilic, it is prone to adsorb more water rather than *p*-chlorophenol, so little p-chlorophenol was adsorbed to the surface of MMT-Na, and the adsorption amount reveal slow growth as the equilibrium concentration increase. After modification, the surface of modified montmorillonite changes from hydrophilic to hydrophobic, the surface energy reduced. With a paraffin structure, the interlayer of the modified montmorillonite exhibited good compatibility with *p*-chlorophenol, adsorption and partition together affect the adsorption of *p*-chlorophenol on the modified montmorillonite, leading to high adsorption to *p*-chlorophenol.

4. Summary

FTIR analysis proves that the Gemini cationic surfactants have been successfully intercalated into the montmorillonite. The TG curve shows that there is little mass loss at 35-200°C regions, while the organic constitutes mainly decomposed in 200-500°C region. The XRD results show that the surfactants used in this paper can expand the basal spacing of the montmorillonite effectively and have more advantages than the other Gemini cationic surfactants. The two intercalators form a paraffin structure in the montmorillonite interlayer. The modified montmorillonite revealed a better removal of p-chlorophenol than the unmodified montmorillonite in adsorption test, which showed a promising application on the removal of the organic pollutes in wastewater.

References

Chen, F.S., Liu, X.M., Hu, Z.R., Zhao, C.X., 2008a. Synthesis and properties of cationic Gemini surfactants with amide group. *China Surfactant Detergent & Cosmetics.* 38 (5) 285-288.

Chen, D., Zhu, J.X., Yuan, P., Yang, S.J., Chen, T.H., He, H.P., 2008b. Preparation and characterization of anion-cation surfactants modified montmorillonite. *J. Therm. Anal. Cal.* 94, 841-848.

He, H.P., Ding, Z., Zhu, J.X., Yuan, P., Xi, Y.F, Yang, D., Frost, R.L., 2005. Thermal characterization of surfactant-modified montmorillonites. *Clays and Clay Minerals.* 53,287-293.

Huskic, M., Brnardic, I., Zigon, M., Ivankovic, M., 2008. Modification of montmorillonite by quaternary polyesters. *Journal of Non-crystalline Solids.* 354, 3326-3331

Joanna, W., Jan, C., 2006, Preparation, Surface-Active and Antielectrostatic Properties of Multiple Quaternary Ammonium Salts. *Journal of Surfactants and Detergents.* 9 (3), 221-226.

Juang, R-S., Lin, S-H., Tsao, K-H., 2004. Sorption of phenols from water in column systems using surfactant-modified montmorillonite. *Journal of Colloid and Interface Science.* 269, 46-52.

Kozak, M., Domka, L., 2004. Adsorption of the quaternary ammonium salts on montmorillonite. *Journal of Physics and Chemistry of solids.* 65, 441-445.

Lagaly, G., 1986. Interaction of alkylamines with different types of layered compounds. *Solid State Ionics.* 22, 43-51.

Lawrence, M.A.M., Kukkadapu, R.K., Boyd, S.A., 1998. Adsorption of phenol and chlorinated phenols from aqueous solution by tetramethylammonium- and etramethylphosphonium-exchanged montmorillonite. *Applied Clay Science.* 13, 13-20.

Li, F., Rosen, M.J., 2000. Adsorption of Gemini and Conventional Cationic Surfactants onto Montmorillonite and the Removal of Some Pollutants by the Clay. *Journal of Colloid and Interface Science.* 224, 265-271.

Pernyeszi, T., Kasteel, R., Witthuhn, B., Klahre, P., Vereecken, H., Klumpp, E., 2006. Organoclays for soil remediation: Adsorption of 2, 4-dichlorophenol on organoclay/aquifer material mixtures studied under static and flow conditions. *Applied Clay Science.* 32, 179-189.

Rosen, M.J., Li, F., 2001. The Adsorption of Gemini and Conventional Surfactants onto Some Soil Solids and the Removal of 2-Naphthol by the Soil Surfaces. *Journal of Colloid and Interface Science.* 234, 418-424.

Wibulswas, R., White, D.A., RAUTIU, R., 1999. Adsorption of phenolic compounds from water by surfactant-modified pillared clays. *Trans IChemE.* 77, 88-92.

Xie, W., Gao, Z.M., Pan, W.P., Hunter, D., Singh, A., Vaia, R., 2001. Thermal Degradation Chemistry of Alkyl Quaternary Ammonium Montmorillonite. *Chem.Mater.* 13, 2979-2990.

Xie, W., Xie, R.C., Pan, W.P., Hunter, D., Bryan,Koene., Tan, L.S., Vaia, R., 2002. Thermal stability of Quaternary phosphonium modified Montmorillonite. *Chem.Mater.* 14, 4837-4845.

Xi, Y., Martens, W., He, H., Frost, R.L., 2005a. Thermogravimetric analysis of organoclays intercalated with the surfactant octadecyltrimethylammonium bromide. *J. Therm. Anal. Cal.*

81,91-97.
Xi, Y.F., Frost, R.L., He, H.P., Kloprogge, T., Bostrom, T., 2005b. Modification of Wyoming Montmorillonite Surfaces Using a Cationic Surfactant. *Langmuir.* 21, 8675-8680.
Zampori, L., Gallo Stampino, P., Dotelli, G., 2009. Adsorption of nitrobenzene and orthochlorophenol on dimethyl ditallowyl montmorillonite: A microstructural and thermodynamic study. *Applied Clay Science.* 42, 605-610.
Zhou, L.M., Chen, H., Jiang, X.H., Lu, F., Zhou, Y.F., Yin, W.M., Ji, X.Y., 2009. Modification of montmorillonite surfaces using a novel class of cationic gemini surfactants. *Journal of Colloid and Interface Science.* 332, 16-21.
Zhu, R.L., Zhu, L.Z., Zhu, J.X., Xu, L.H., 2008. Structure of cetyltrimethylammonium intercalated hydrobiotite. *Applied Clay Science.* 42, 224-231.

Nano-Structured Sorbents for Desulfurization of Biomass-Derived Syngas

Mayank Behl[1], Junghoon Yeom[2] and Mark A. Shannon[2],
[1]Chemical and Biomolecular Engineering, University of Illinois at Urbana-Champaign, Urbana, IL,
[2]Mechanical Science and Engineering, University of Illinois at Urbana-Champaign, Urbana, IL

Abstract

Nanofibrous mats made up of composite zinc and titanium oxides have been synthesized using the method of electrospinning for their application as regenerable sorbents for high temperature desulfurization of biomass-derived syngas. For obtaining the information related to the surface morphology, specific surface area, composition and the crystal structure, fiber mats have been characterized using the techniques of SEM, EDX, BET & X-ray diffraction. Thermo-gravimetric analysis has also been carried out in order to study sorbent's response when subjected to reducing atmosphere. Comparison with the data available on bulk zinc titanate adsorbents points to fibers' enhanced resistance to reduction.

1. Introduction

Production of syngas via gasification of biomass residue has gained considerable attention worldwide. Non-edible agriculture residue, residue from saw and paper mills and forest wood residues constitute majority of the biomass residue. Gasification of biomass, a thermochemical conversion, is particularly attractive as it can operate with a wide range in the quality and consistency of the feedstock, thus increasing the available feedstock and reducing the overall cost. Depolymerizing the source biomass right down to the useful gaseous products like syngas makes the process of gasification the only currently established technology that can claim to extract most of the energy stored in the raw biomass (Seitarides et al., 2008).

However, on gasification, along with the syngas, gaseous impurities like sulfur containing species (H_2S, COS), ammonia, alkali oxides, halides etc are also generated due to the volatile contaminants present in the biomass residue. Many downstream processes, like Fischer Tropsch synthesis, Solid Oxide Fuel Cells and methanol production, use catalysts that have little tolerance with gaseous contaminants. Transition metal / metal oxide catalysts, typically employed for such value-addition processes, are especially vulnerable to sulfur containing gaseous species like hydrogen sulfide. Therefore, before this raw syngas can be used for different downstream applications, it needs to be cleaned. Among the different processes used for H_2S removal, wet amine scrubbing has been the most popular. However, such low temperature processes require comparatively low operating temperatures (35 - 55^oC). The cleaned syngas has to be subsequently reheated for the downstream processes (300 - 800^oC). Such consecutive cooling and heating can cause considerable thermal losses.

To avoid such energy losses, it is important to desulfurize the gas stream at suitably high temperatures. A solid-phase sulfur-sorbent material that has a high reactivity, good structural stability, and easy regenerability at such high temperatures can provide such an alternative. In past work, different types of bulk sorbents with different chemical and structural properties have been tested for the applications involving high temperature desulfurization, pertaining mainly to the research in the field of coal gasification. Zinc oxide happens to be one of the most extensively studied metal oxides used for desulfurization applications. Zinc oxide is known for its high sulfidation capacity, favorable thermodynamics and high efficiency when it comes to sulfur removal. However, as shown by Harrison (1998), ZnO starts getting reduced to zinc metal at high operating temperatures (>550 C) due to the presence of gases like H2 and CO in the exit gas stream of a gasifier. Thus, the use of ZnO is limited to only few-cycles at mid-high temperatures. In order to develop sorbents that are suited to a multicycle operation, other metal oxides like Fe2O3 and TiO2 have frequently been used with ZnO to synthesize mixed metal oxides with an aim of stabilizing zinc oxide against reduction. Zinc ferrites suffer from similar problems as at high temperatures they tend to decompose to individual oxide components, which subsequently reduce towards their metal form. Zinc titanates, however, have been shown to be much more effective against reduction. When compared to ZnO, Lew et al. (1989, 1992) showed that the rate of reduction was about five to ten times slower for sorbents containing Zn-Ti-O. However, although the presence of Titania does decrease the reduction tendency of the Zn-Ti-O system, it also leads to lower overall conversion and decrease in the specific capacity of the sorbent, mainly due to the fact that titania does not take part in the reaction.

In short, it can be said that in spite of numerous efforts to modify chemical and compositional properties, only a limited success has been achieved. It is primarily due to sorbent's failure in meeting one or more of the above-mentioned criteria. In addition, it has been demonstrated that due to mass transfer limitations (Focht et. al, 1989), if a bulk sorbent gets completely sulfided, it is almost impossible to achieve complete regeneration afterwards, regardless of process's favorable thermodynamics and chemical kinetics. This incomplete regeneration, partly due to mass transfer limitation and subsequent loss of sorbent's surface area, leads to underutilization of the sorbent material, which is typically a transition or rare earth metal oxide. To overcome these limitations, it seems necessary that one needs to go beyond doing modifications in sorbent chemical composition.

One option that has been little explored involves investigation of the performance of structured sorbent materials having microstructures and surface morphologies that can be tailored to suit the process. Integration of such capability with the material's inherent favorable thermodynamics and kinetics can lead to new insights. Controlling and characterizing these properties is difficult at the scale of bulk sorbents. Structuring and nanosizing of the sorbents can allow for such changes much easily. The approach, therefore, should be to come up with a sorbent design that facilitates short contact time so that deep sulfidation can be avoided without affecting the sulfur removal capacity. Moreover, the intrinsic reaction rate has been shown to be comparatively short (Wang & Stephanopoulos, 2005). As short contact time necessitates the use of high gas velocity, the use of conventional bulk sorbent will not only lead to significant pressure drop but will also lead to underutilization of the sorbent material. Sorbents with tailored designs and sizes, however, can help in overcoming such limitations by providing considerably higher specific surface area and short diffusion lengths for minimizing the mass transfer resistances. Nanosizing of sorbent seems as a good alternative. However, sorbent in the form of nano-particles tend to aggregate and may cause diffusion limitations similar to bulk sorbents. Dispersion of the particles on a support structure seems to be another option; however, being adsorbent (and not catalyst), they will be needed in comparatively larger quantities and catalyst-like dispersion on a support surface may limit the sulfur removal capacity. Therefore, what is needed is a synthesis technique that can produce nanoscale sorbent structures that do not tend to aggregate and at the same time is amenable to scale up to bulk-like production levels. Ceramic nanofibers, prepared using the process of electrospinning, have shown such potential (Dai et al., 2010). This process can be used to synthesize high surface area metal oxide fibers with diameter as small as 50 nm. Because of their high aspect ratio, these nanofibers tend to remain isolated; thus, potentially providing a more suitable framework for carrying out frequent cyclic sulfidation-regeneration operation. Such high aspect ratio nanoscale structures not only retain the favorable thermodynamics and chemical affinity of their bulk equivalents but also develop useful properties due to anisotropic geometry and reduced grain sizes. Given that the reactions involved are inherently fast, abovementioned properties of the nanofibers, along with the operating conditions that facilitate short contacting time, can help in limiting the excessive volumetric changes that typically result in sorbent failure in cyclic sulfidation/regeneration operation. Thus, it is expected that the use of sorbents in the form of stacked fiber-mats made from such nanofibers can not only lead to considerably higher specific surface area and improved mass-transfer, they may also lead to an improved mechanical integrity during high temperature cyclic operation.

The setup for electrospinning consists of three major components: a metallic needle, high voltage source and a metallic collector. The metallic needle is called spinneret. It is attached to syringe that contains a polymeric solution homogenously mixed with a salt or a propoxide solution of the given metal. A syringe pump is used to feed this mixture through the spinneret at a controllable rate. When a high voltage is applied (typically via DC power), the pendant drop at the tip of the metallic needle becomes highly charged with charges induced all over the surface of the drop. In addition to the solution's own surface tension, drop also experiences two other forces: electrostatic repulsion due to surface charges and Coulombic force exerted by the external electric field. Under the action of these forces, the shape of the drop will distort from spherical to conical, commonly known as Taylor cone. Once the strength of the electric field exceeds the surface tension of the solution, a liquid jet will be ejected from the nozzle (Li & Xia, 2004). This electrified jet then undergoes stretching and whipping process, leading to a long, thin thread. Solvent evaporates as the jet continuously elongates, and in the process, the thread's diameter starts reducing by several orders. This charged fiber is deposited on the grounded collector as a randomly oriented, non woven web or mat. This fiber-mat is first vacuum dried and then sintered at elevated temperatures, removing all the organic compounds present in the green fibers. Ramaseshan et al. (2007) recently used electrospinning to successfully produce ZnO/TiO2 fibers. They investigated these fibers for their reactivity against chemical warfare agents. They reported satisfactory performance in detoxifying CEES (2-chloroethyl ethyl sulfide) which was being used as a liquid phase simulant for mustard agents.

In the present work, high surface area composite Zinc and titanium oxide fibers were synthesized using the process of electrospinning. Fiber mats were characterized using the techniques of BET, SEM and XRD. The aim is to investigate such fibrous sorbents for their potential to be used as an alternative adsorbent design for multi-cycle desulfurization applications.

2. Experimental

2.1. Materials

A sol-gel based approach was used to prepare the precursor solution for electrospinning. Zinc acetate dehydrate 97% {CH3COO}2Zn.2H2O (Alfa Aesar, CAS #5970-45-6) and Titanium (IV) isopropoxide 97%, C12H28O4Ti (Sigma Aldrich, 97%, CAS#546-68-9) were used as the starting material. Ethanol (ACS grade) was used as the solvent and glacial acetic acid was used to stabilize the solution. In order to obtain specific Zinc to Titanium ratio in the final sorbent, the relative amounts of the zinc salt and the titanium isopropoxide should be changed accordingly during this step. These chemicals were mixed with ethanol and acetic acid until a clear solution was obtained. This ceramic precursor solution was then mixed with a polymeric solution prepared using PVP (Mw ~ 1300000, Sigma-Aldrich, CAS#9003-39-8) dissolved in ethanol. Table 1 shows quantities of different chemicals used for one such solution.

Table -1: Chemicals used for the Sol-gel based Zinc Titanate fibers

Component	**Amount**
$Zn(CH3COO)_2.2H_2O$	5.6 g
$Ti(O\text{-}iPr)_4$	4.0 ml
Acetic acid	9.0 ml
PVP	2.4 g
Ethanol (net)	22.0 ml

2.2. Sorbent preparation: Metal oxide Fiber synthesis using Electrospinning

The ceramic precursor and polymer solution were mixed for 6hrs at 55^{0}C. The solution thus obtained was drawn into a 10 ml syringe. Syringe was equipped with a 20 gauge blunt metal needle. An aluminum mesh was used as the collector. A voltage of 19kV was applied to the needle while the collector was grounded. The distance between the needle tip and the collector was fixed at 17 cm. The feed rate was fixed at 1.2 ml/hr using a syringe pump. Fig-1 shows the picture of the setup.

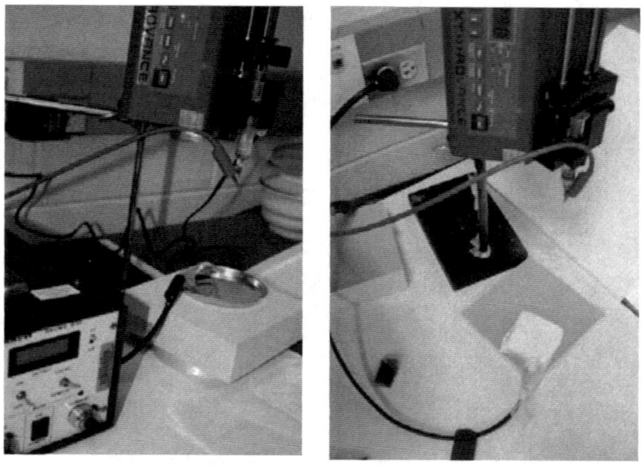

Figure-1: Pictures of the electrospinning setup

The composite inorganic-organic fiber mats (referred to as green or as-spun fibers) thus obtained were vacuum dried for 3hrs at 110^{0}C. Afterwards, they were moved to a furnace and subsequently heated to 340C at a rate of 4C/min. Fiber mats were kept at that temperature for 2hrs to remove the polymer and/or any other organic components. Thereafter, the temperature was raised at 2C/min to 600^{0}C. The fibers were sintered at 600^{0}C for 4hrs in air. Depending upon the ratio of Zn to Ti in the precursor solution, fiber mats with different composition were obtained.

2.3. Results from Sorbent Characterization

2.3.1. X-Ray Diffraction

Technique of X-Ray Diffraction was used to identify the different crystalline phases present in the sintered fibers. An XRD instrument (X'pert MPD Pro, Philips®) with Cu Kα radiation (λ = 1.54 Å) was used for this analysis. The XRD spectrum generated by one such fiber mat (Sample-1) is shown in Fig-2. All the major peaks were identified and assigned to the corresponding crystal phases in accordance with JCPDS database.

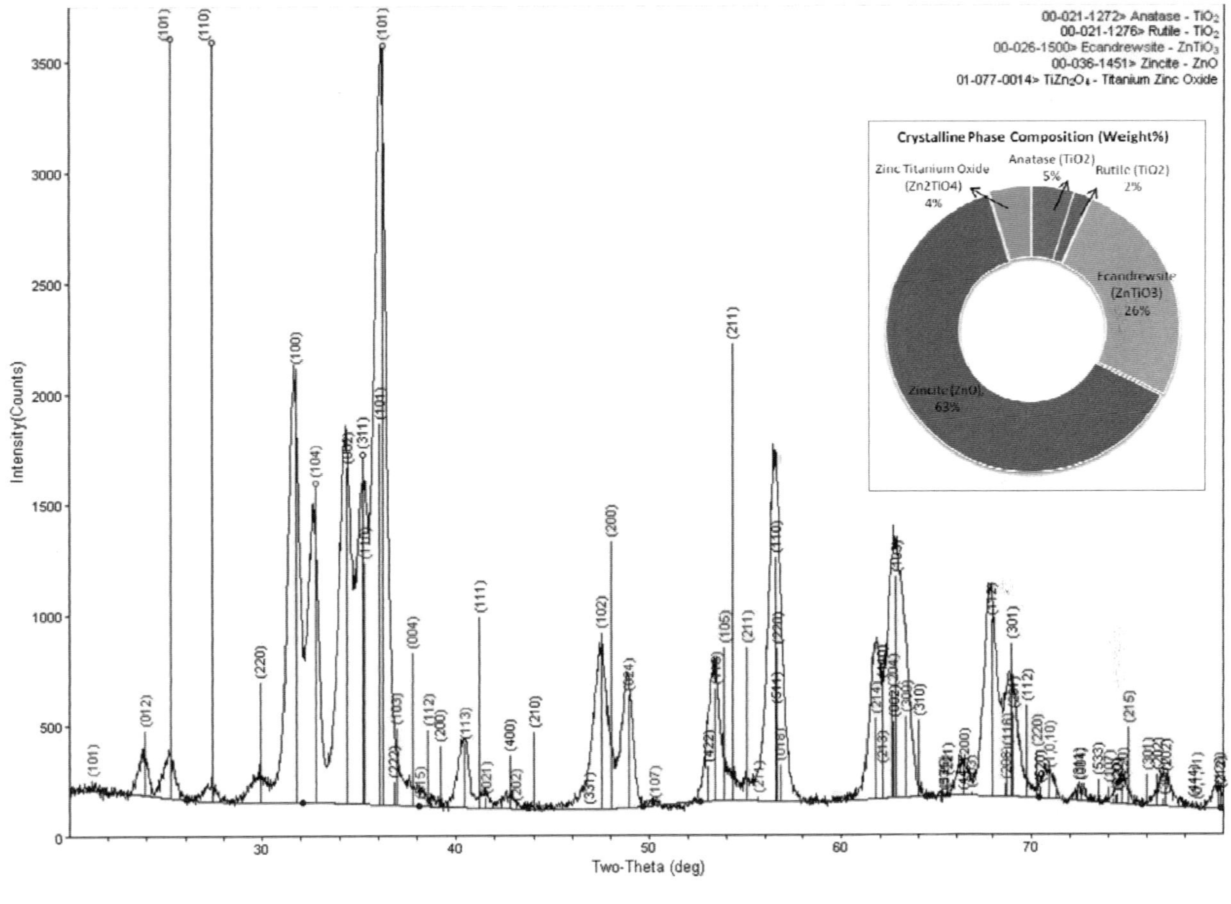

Figure-2: X-ray diffraction pattern of ZnO- TiO_2 nanofibers (Sample-1)

The major phases identified in one of the sample were: $ZnTiO_3$ (Ecandrewsite), ZnO (Zincite), TiO_2 (Rutile, Anatase) and minimal amount of Zn_2TiO_4. Fig-2 (Inset) also shows the relative amount of different crystalline phases identified in the Sample-1 (one of the calcined fibers mats). The relative amount of the different crystalline phases present in a given fiber mat sample (weight basis) were obtained by analyzing profile-fitted peaks (using a software called MDI-Jade®). The d-spacing for some of the atomic planes and the corresponding grain size for the different crystal structures are given in Table-2.

Table-2: The d-spacing and grain size for some of the atomic planes of different phases (Sample-1)

Chemical compound/phase (plane)	d-spacing (nm)	grain size (nm)
$ZnTiO_3$ (012)	0.372	15.0
TiO_2 anatase (101)	0.353	13.0
TiO_2 rutile (110)	0.326	10.9
Zn_2TiO_4 (220)	0.299	9.0
ZnO (100)	0.282	13.1
$ZnTiO_3$ (104)	0.273	15.1
ZnO (112)	0.138	11.6

Fiber mats with different compositions (Zn-to-Ti ratio) were synthesized. In order to illustrate the effect of composition, characterization results from a second sample (Sample-2a) are also discussed here. Fig-3 compares the crystalline phase compositions (weight percentage) for the two different mats: Sample-1 & Sample-2a. These two fiber-mats were selected as the model samples for studying the effect of sorbent-composition. Table-3 compares the Zn-to-Ti ratio (atomic percentage) and other important characteristics of these two samples.

Table-3: Comparison of the properties of two different samples

Sorbent (Fresh)	Zn-to-Ti atomic ratio	Surface Area, α_0 (m^2/g)	Mean Fiber diameter (nm)	Avg. grain size (nm)
Sample-1	3.69	151.7	435	12.53
Sample-2a	1.17	90.1	714	15.82

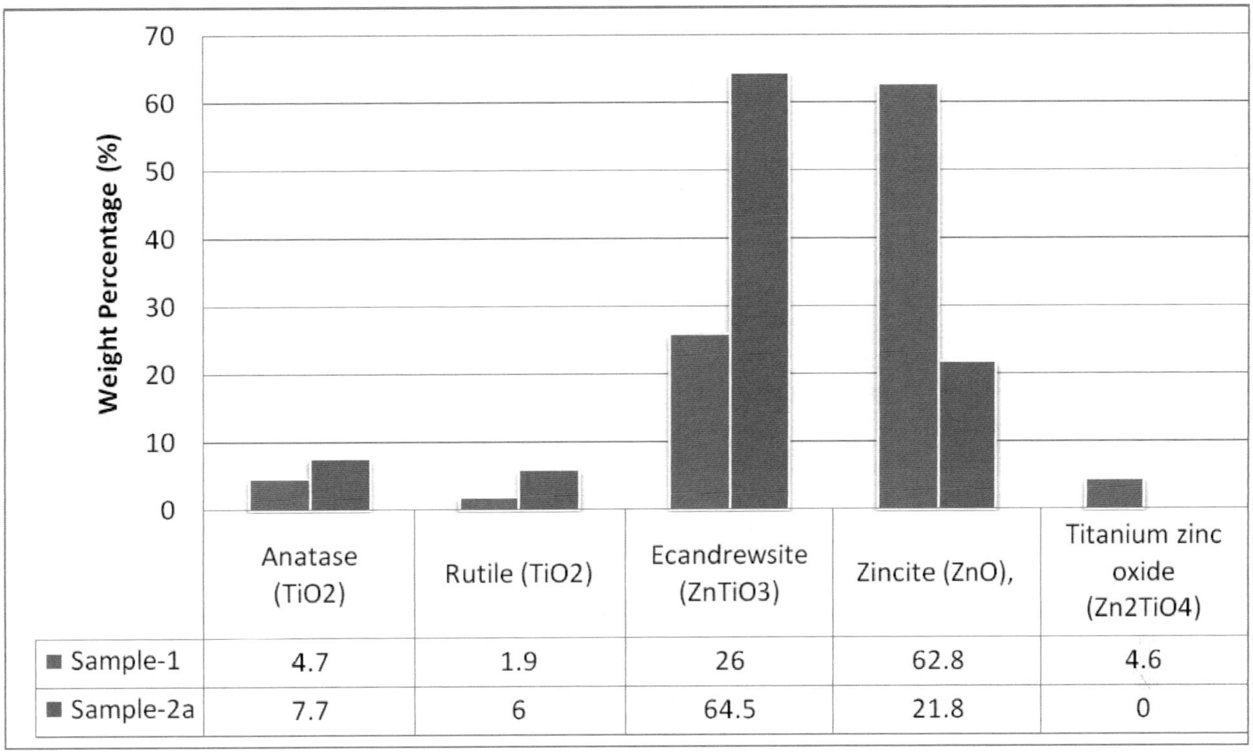

	Anatase (TiO2)	Rutile (TiO2)	Ecandrewsite (ZnTiO3)	Zincite (ZnO),	Titanium zinc oxide (Zn2TiO4)
Sample-1	4.7	1.9	26	62.8	4.6
Sample-2a	7.7	6	64.5	21.8	0

Fig-3: Comparison of the composition of two different samples

2.3.2. Electron microscopy

Surface morphology of the fiber-mats was studied using the technique of Scanning Electron Microscopy (SEM). A JEOL 6060LV® instrument was used for this purpose. Fig- 4 shows the SEM image of the calcined fibers obtained from Sample-1 & 2a. Also shown in the figure is the fiber size distribution corresponding to the selected samples. An image analysis software (Image-J®) was used to estimate the size distribution.

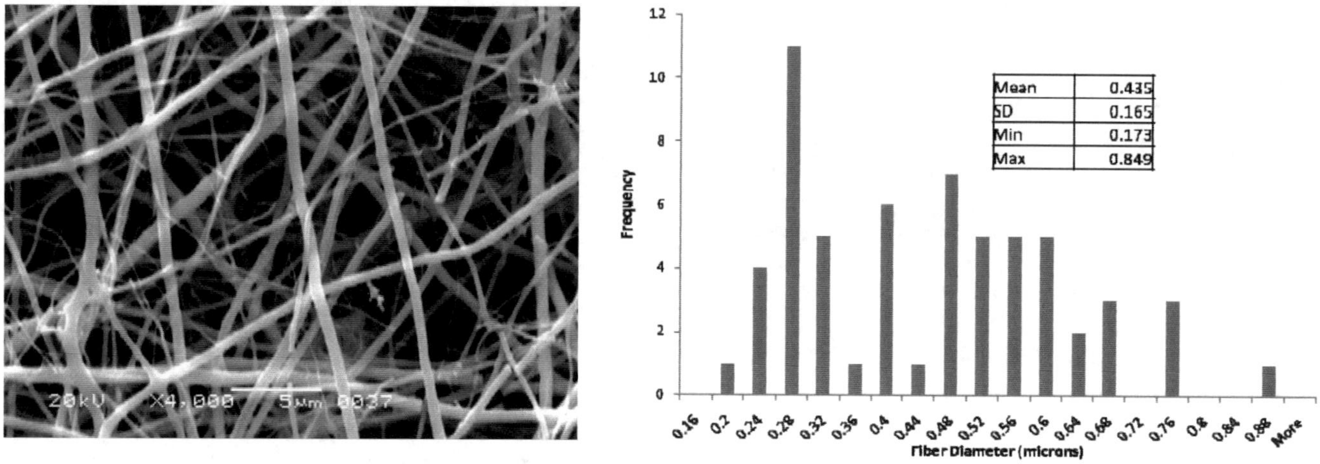

4a)

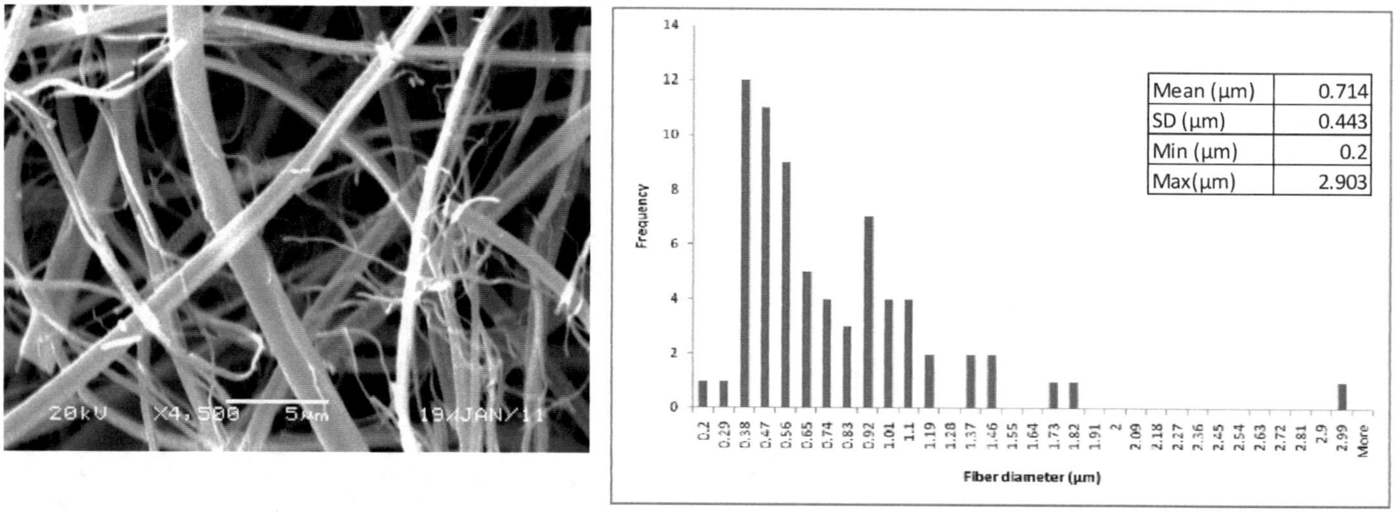

4b)

Figure-4: SEM of Zinc Titanate fiber mats: a) Sample-1, b) Sample-2a

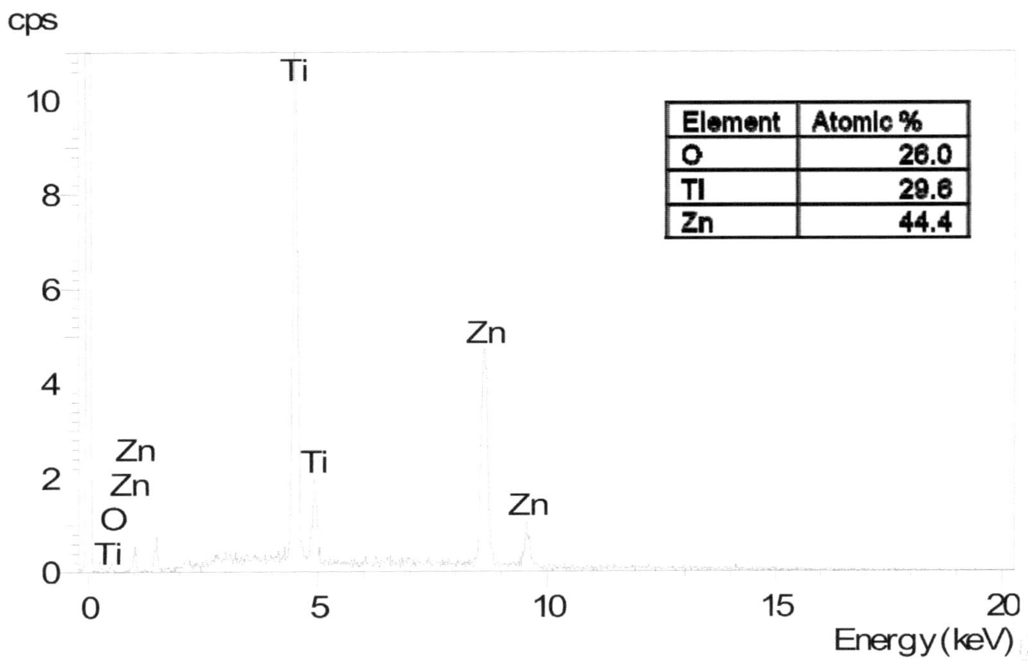

Figure-5: EDX spectra of Zinc Titanate fiber mat (Sample-2a)

Energy-Dispersive X-Ray analysis (EDX) was also conducted for semi-quantitative estimation of the elemental composition of the portion of the sample being imaged. The setup for EDX was integrated with the SEM instrument. Fig- 5 shows the spectra obtained for Sample-2a along with atomic composition of the image.

2.3.3. Surface area measurements

BET specific surface area (α_0) of the sintered fibers was measured using a surface area analyzer (Quantachrome, Nova 2200 Series). It was found to be 151.7 m^2/g for Sample-1 and 90.1 m^2/g for Sample-2a.

2.3.4 Estimation of Reduction Kinetics via Thermo-Gravimetric Analysis (TGA)

For most of the applications that involve sorbents for desulfurization, reduction reaction occurs simultaneously along with the sulfidation reaction. In presence of gases like H_2 and CO, most of the metal oxides tend to get reduced to their lower oxides form (Fe_2O_3 to FeO) or worse, to their metal form (ZnO to Zn). Early work on hot gas desulfurization (Harrison & Atimtay, 1998; Lew et al., 1989), have confirmed that the lower oxide forms, for majority of the oxides, have inferior sulfidation thermodynamics and kinetics (exception: Wang & Stephanopoulos (2005) found that Ce_2O_3 works better than CeO_2). This is one of the major reasons that affects sorbent's lifetime adversely and limits its application to few cycles only. Sorbent needs to be selected and produced in such a way so as to offer maximum possible resistance to the reduction reaction. Therefore, for the present study, information obtained from the investigation of sorbents' reduction kinetics was considered important in determining its suitability to serve as a high temperature sorbent for sulfur removal. For performing these tests, a thermo-balance (Cahn's TGA, Thermax-500) was used as an isothermal differential reactor. A small, known quantity of fibers (~37.16 mg (W0), Sample-1) was placed inside the TGA's quartz container. The temperature was raised to the desired temperature (600°C) at the rate of 15°C/min. During this period only Nitrogen was flowing through the balance. Once the selected temperature was attained, a gas stream containing 5% H_2 gas (rest N_2) was introduced at a total flow rate of 100 ml/min. The flow rate was controlled by a gas mass flow meter integrated with the TGA's temperature controller. The reaction was allowed to proceed for approximately 140 minutes.

Decrease in the sample weight was attributed to the following reduction reaction:

$$Zn_xTi_yO_{x+2y(s)} + x.H_{2(g)} \Leftrightarrow x.Zn_{(g)} + x.H_2O_{(g)} + y.TiO_{2(s)}$$

It was assumed that only zinc oxide gets reduced at 600°C. It has been reported earlier that titania does not suffer any significant reduction at temperatures below 1000°C (Lew et al., 1992).

Fig- 6 shows the change in weight with time as recorded by the software. The initial rate of reduction was defined as flowing:

$$R_0 = \frac{(dW/dt)}{A_0 \cdot M_{ZnO}}$$

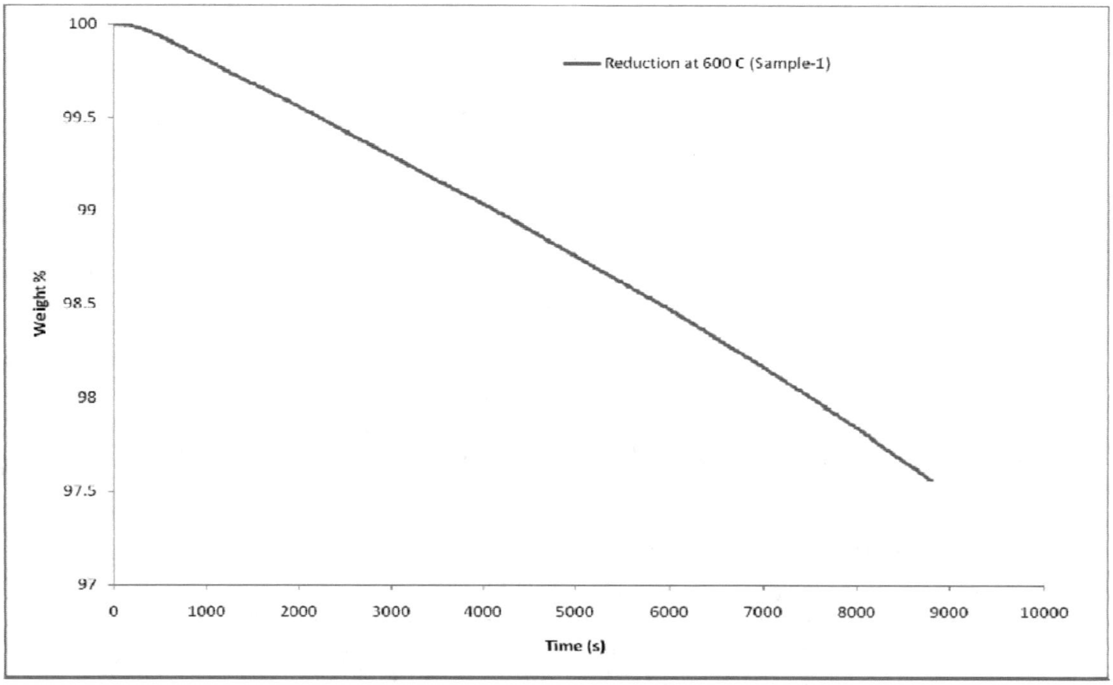

Figure-6: Gravimetric analysis using TGA (Sample-1)

Where
dW/dt = initial rate of change in the sample weight (attributed to loss of Zinc metal ($Zn_{(g)}$) and oxygen atom taken up in $H_2O_{(g)}$)
A0 = Total area of the sorbent at t = 0 ($A_0 = W_0{*}\alpha_0$)
M_{ZnO} = Molecular weight of Zinc Oxide
Conversion (X) was defined as following:

$$X = \frac{\Delta W}{C_0.(M_w - \frac{y}{x}M_{TiO2})} \times 100$$

Where $C_0 = W_0 / M_w$
x, y= atomic fraction corresponding to Zinc and Titanium (estimated by quantitative analysis of results obtained from XRD or EDX (XRD used in present case assuming that all of the sample is crystalline).
M_w = Effective molecular weight of the sample (calculated using y, x)

Results from Sample-1 are shown in Table-4

Table-4: Estimation of initial reduction rate and overall conversion

Sorbent	**Zn-to-Ti atomic ratio (x/y)**	**Effective Molecular weight (M_w)**	**Surface Area, α_0 (m^2/g)**	**W_0 (mg)**	**Conversion X (%)**	**Initial Rate of reduction, R_0 ($moles/min/m^2$)**
Sample-1	3.69	96.78 g/mol	151.7	37.17	3.1	1.36E-08

Initial rate of reduction (R_0) for the nanofibrous-mat comes out to be 1.36E-08 Moles/min/m^2 or 2.26E-11 mMoles/s/cm^2. It is approximately three orders of magnitude smaller than the value reported for the bulk Zinc titanate adsorbents studied by Lew et al. (1992) with similar operating conditions. Reduction rate of around 6E-8 mMoles/s/cm^2 was reported at 600°C for a sorbent (Z3T) having Zn-to-Ti atomic ratio (x/y) of 3:1. This result points to the role played by a composite sorbent's surface morphology and grain size. The presence of much smaller grain size (≈ 10 - 15nm from Table-2) in the sub-micron fibers, is apparently leading to much better dispersion of titanium that is lowering the reduction rate possibly by more effective elimination of sites for hydrogen dissociation than that is possible in the bulk sorbents. A set of reduction experiments occurring at different temperatures have been planned as the future work for estimating the activation energy and reduction mechanism associated with the proposed sorbent structure.

3. Summary and conclusion

Process of electrospinning was used to synthesize composite zinc and titanium oxide adsorbents in the form of non-woven fibrous mats. Sol-gels, with a desired proportion of Zinc-to-titanium, were first prepared as a precursor solution. Calcination of the as-spun fibers at 600°C resulted in zinc-titanate fibers free from polymer. Fibers were characterized by determining the crystal structure and surface morphology using SEM and XRD, and were subsequently tested for their high temperature reduction behavior, considered important to any cyclic sulfidation-regeneration operation. Fiber-based sorbents offered significant improvement in resistance to reduction. In future, experiments studying the differences in the sorbent performance after each sulfidation cycle should be analyzed. The results from such investigation may help in estimating the different ways in such nanostructured sorbents can prove to be useful during a multi-cycle sulfidation operation.

Acknowledgement

Sample characterization was carried out in part in the Frederick Seitz Materials Research Laboratory's Central Facilities, University of Illinois, which are partially supported by the U.S. Department of Energy.

References

1. Cheah, S., Carpenter, D. L., & Magrini-Bair, K. (2009). Review of mid- to high-temperature sulfur sorbents for desulfurization of biomass- and coal-derived syngas. Energy & Fuels, 23(11), 5291-5307.
2. Dai, Y., Liu, W., Formo, E., Sun, Y., & Xia, Y. (2010). Ceramic nanofibers fabricated by electrospinning and their applications in catalysis, environmental science, and energy technology. Polymers for Advanced Technologies
3. Focht, G.D., Sa, L.N., Ranade, P.V., Harrison, D.P. (1989). High-temperature desulfurization using zinc ferrite: Solid structural property changes. *Chemical Engineering Science, 44*(2), 215-224
4. Flytzani-Stephanopoulos, M.; Li, Z (1998). Kinetics of sulfidation reactions between H2S and bulk oxide sorbents. In Desulfurization of Hot Coal Gas, Atimtay, A. T.; Harrison, D. P., Springer: Berlin, Vol. 42, pp 179-211.
5. Harrison, D. P. (1998). Performance analysis of ZnO-based sorbents in removal of H2S from fuel gas. In Desulfurization of Hot Coal Gas, Atimtay, A. T.; Harrison, D. P., Springer: Berlin, Vol. 42, pp 213-242.
6. Li, D., & Xia, Y. (2004). Electrospinning of nanofibers: Reinventing the wheel? Advanced Materials, 16(14), 1151-1170.
7. Lew, S., Jothimurugesan, K., & Flytzani-Stephanopoulos, M. (1989). High-temperature hydrogen sulfide removal from fuel gases by regenerable zinc oxide-titanium dioxide sorbents. *Industrial & Engineering Chemistry Research, 28*(5), 535-541.
8. Lew, S., Sarofim, A. F., & Flytzani-Stephanopoulos, M. (1992). The reduction of zinc titanate and zinc oxide solids. *Chemical Engineering Science, 47*(6), 1421-1431.
9. Li, D., McCann, J. T., Xia, Y., & Marquez, M. (2006). Electrospinning: A simple and versatile technique for producing ceramic nanofibers and nanotubes. Journal of the American Ceramic Society, 89(6), 1861-1869.
10. Ramaseshan, R. and Ramakrishna, S. (2007). Zinc titanate nanofibers for the detoxification of chemical warfare simulants. J.Am.Ceram.Soc, 90(6), 1836-1842.
11. Ramaseshan, R., Sundarrajan, S., Jose, R., & Ramakrishna, S. (2007). Nanostructured ceramics by electrospinning.102(11), 17.
12. Seitarides, T., Athanasiou, C., & Zabaniotou, A. (2008). Modular biomass gasification-based solid oxide fuel cells (SOFC) for sustainable development. Renewable and Sustainable Energy Reviews,12(5), 1251-1276.
13. Wang, Z., & Flytzani-Stephanopoulos, M. (2005). Cerium oxide-based sorbents for regenerative hot reformate gas desulfurization. Energy & Fuels, 19(5), 2089-2097.

(55b) Standing Wave Design of Simulated Moving Beds for the Recovery of Sugars From Biomass Hydrolysate

Nien-Hwa Linda Wang, School of Chemical Engineering, Purdue University, West Lafayette, IN

Simulated Moving Bed (SMB) is more economical than batch adsorption for large scale purification. They have been used to produce para-xylene since 1970's, high-fructose corn syrup since 1980's, and chiral pharmaceuticals since 1990's. However, SMB has not been widely used for multicomponent separation. A major barrier is the difficulty in designing zone flow rates and port velocity to ensure high purity and high yield in systems with significant diffusion effects. SMB has three or more zones, which have different zone flow rates. Design the flow rates and port velocity by trial and error using either experiments or process simulations will require a large number of trials. A second barrier is the lack of design method to control the split of a mixture of three or more components. To overcome these barriers, a design method based on the concept of standing concentration waves in SMB was developed. A difference between port velocity and wave velocity is designed to focus a key wave in each zone to control product purity, yield, and the split. When intrinsic adsorption and diffusion parameters are known, the necessary flow rates and step times can be found quickly using minimal calculations. This method also reduces optimization search space by four dimensions or more. This talk will discuss how this method was used to develop a nine-zone SMB to recover glucose and xylose from a three component mixture and a five-zone SMB to recover six sugars with high purity (>95%) and high yield (>99%) from biomass hydrolysate (a complex mixture with 10 components).

See more of this Session: University Research In the Field of Biomass/Renewables
See more of this Group/Topical: Topical 5: University Research for Industry

(55c) Sustainable Polymeric Materials Prepared by Ultrafast Photo-Polymerization of Chemically Modified Vegetable Oils

Na Yeon Kang, Chemical & Biochemical Engineering, University of Iowa, Iowa City, IA, Brian Dillman, Department of Chemical & Biochemical Engineering, The University of Iowa, Iowa City, IA and Julie L. Jessop, University of Iowa, Iowa City, IA

Photopolymerization is a technique that uses light rather than heat to initiate polymerization reactions. Photopolymerization has been used extensively in adhesives, coatings, and lithographic industries due to the high curing rates and spatial and temporal control realizable with it. While conventional petroleum-based monomer feeds have been used in a highly energy efficient manner through photopolymerization, the development of polymeric materials based on renewable resources is critical to sustainability efforts. In this work, vegetable oils were used as base materials for the fabrication of polymeric films by several photo-induced polymerization mechanisms. Using real-time Raman spectroscopy, raw vegetable oils were shown not to undergo rapid photopolymerization at room temperature. Thus, vegetable oils were chemically modified in order to increase reactivity by the addition of epoxide, (meth)acrylate, and norbornene moieties to the triglyceride molecules. Modified oils were then polymerizable by radical, anionic, or cationic mechanisms depending on the functional groups attached to the oil substrate. The modified oils were photo-polymerized at room temperature, and quantitative conversions were achieved. The physical/mechanical properties of the materials produced varied widely depending on the functional groups attached to the triglyceride and the polymerization mechanism used. The glass transition temperature of polymer products was easily modified in both the breadth of the transition and the temperature at which the transition occurs by the blending of various modified oils with conventional monomers. Additionally, the cured materials were subjected to accelerated aging by high-temperature storage, simulated sunlight, and combinations of both. The property development and stability of mechanical performance were correlated to the additives used to cure the materials.

See more of this Session: University Research In the Field of Biomass/Renewables
See more of this Group/Topical: Topical 5: University Research for Industry

Syntheses of Gemini Betaine Surfactants from α- Chlorocarboxylic Acids

Guangming Du[1, 2], *Yun Fang*[1,*]
1. School of Chemical & Material Engineering, Jiangnan University, Wuxi, Jiangsu 214122, China
2. College of Chemical Engineering, Xinjiang Agricultural University, Urumqi, Xinjiang 830052, China

Abstract

A novel Gemini betaine surfactant is synthesized from α-chlorocarboxylic acids by amination with dimethylamine in this paper, which is a simple and economical process with industrial prospects since the intermediate α-chlorocarboxylic acid is derived from natural and cheap biomass raw material fatty acids through selectively chlorinating by molecular chlorine at the α-position of fatty acids.

Keywords: Gemini surfactant, Zwitterionic surfactant, Betaine, Fatty acid.

Introduction

Dimeric and oligomeric surfactants are increasingly attracting interest in the academic and industrial fields on surfactants owing to their unique properties [1]. Recently the synthesis and the study on the applications and properties of Gemini surfactants has become a scientific topic [1-5]. Some cationic Gemini surfactants have been synthesized due to their easy preparation [6-8]. Syntheses of anionic and nonionic Gemini surfactants are somewhat more complicated [9], and the literature contains relatively few systematic studies of their physico- chemical behavior. However, in comparison with other classes of Gemini surfactants, the difficulty in synthesis of zwitterionic Gemini surfactants is great, thus there have been few reports on this class of surfactants till now.

Menger et al. [10] reported a zwitterionic Gemini surfactant in which one ionic group is negatively charged (a phosphodiester) and the other is cationic (a quaternary ammonium salt). The two head groups were separated by a pair of methylene groups.

Alami et al. [11] published a review of developments in heterogemini surfactants. Yoshimura et al. [12,13] synthesized sulfobetaine-type zwitterionic Gemini surfactants and heterogemini zwitterionic surfactants containing ammonium and carboxylate head groups.

A simple method for synthesizing alkylbetaine zwitterionic Gemini surfactants based on 1,2-bis[N-methyl-N-carboxymethyl-alkylammonium] ethane has been also reported by Yujun Feng et al [1].

Erkki Paatero has studied that carboxylic acids can be selectively α-chlorinated with chlorine in the presence of molecular oxygen as a radical trapper and chlorosulfonic acid as an enolizing agent: the only products were the α-monochloro- and α, α-dichlorocarboxylic acids. [14]

[*] Corresponding author: yunfang@126.com.

In this paper we offered a synthesis method of the novel Gemini betaine surfactants from α- chlorocarboxylic acids by amination with dimethylamine. The intermediate α-chlorocarboxylic acids were synthesized at first from natural fatty acids and molecular chlorine through selectively chlorinating at the α-position of fatty acids. Then dimethylamine was reacted with excess of this intermediate in a hermetic reactor, and the product was obtained (Scheme 1). The synthesis route is a simple and economical process with industrial prospects due to using cheap biomass raw material natural fatty acids.

Scheme 1 Synthesis route for gemini surfactants bis (α-carboxyl-α-decyl) tetramethylammonium, BCDTA

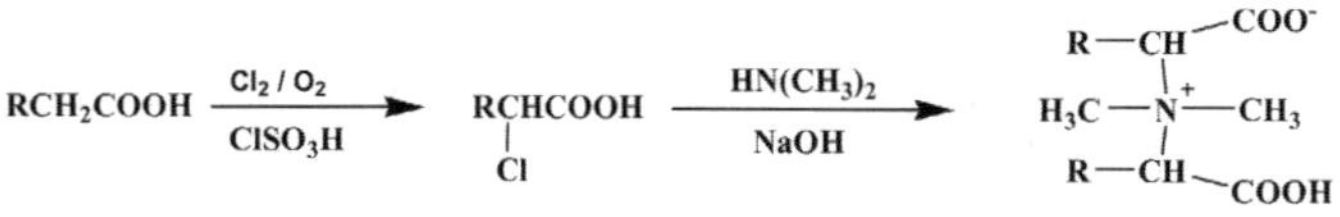

Experimental

Dodecanoic acid (Sinopharm Chemical Reagent Co. Ltd., 99% purity, Shanghai, China), Chlorosulfuric acid (Sinopharm Chemical Reagent Co. Ltd., 99% purity, Shanghai, China) and NaOH (Sinopharm Chemcial Reagent Co. Ltd., analytical reagent grade) were used without further purification. Methanol, ethanol, acetone, ethyl ether were obtained commercially (Sinopharm Chemical Reagent Co. Ltd., 99% purity, Shanghai, China) and used as received.

Synthesis of α-chlorododecanoic acid

The intermediate α-chlorododecanoic acid was synthesized according to the pathway described in first step of Scheme 1, and full details are provided in Ref. [15,16]. The synthesized α-chlorododecanoic acid was refined with petroleum ether(30 ~ 60°C) several times and clear crystals were obtained. The product was dried in a vacuum system overnight and then stored in a vacuum desiccator until it was used. The samples of intermediate were methylated with boron trifluoride-methanol , then the purity (>99%) was verified by capillary GC-FID (FULI 9790, Fuli Analytical Instrument CO., Ltd.) .

Synthesis of bis (α-carboxyl-α-decyl) tetramethylammonium

The ammonolysis reaction was carried out in a hermetic reaction kettle. The intermediate α-chlorododecanoic acid(0.03mol,7.04g) were dissolved in methanol (4.0g). NaOH(0.045mol, 1.80g) was then added to the solution. The mixture was kept at 70 °C in an autoclave with continuous stirring for 2 hours, then cooled to room temperature and dimethylamine (0.01mol, 0.45g) was added to the mixture. The reactor was still kept at 70 °C with stirring for 6 hours. Afterwards, the temperature was increased to 95 °C. The reaction proceeded for 6 h at 95 °C, the mixture then was cooled to room temperature. After the reaction, dimethylamine was evaporated under reduced pressure. The residual α-chlorododecanic acid was extracted with ethyl ether. After the residual mixture was dried, the residue was treated with ethanol. Insoluble salt was filtered off. The filtrate was crystallized. The crude product was thus obtained in powder form.

Analyses of the product

The purity of the product was confirmed by high performance liquid chromatograph (HPLC). The samples of refined product was resolved in methanol and then analyzed by HPLC. The HPLC separation was carried out on a Waters 1525 Binary HPLC Pump (Waters Analytical Instrument Co., Ltd., USA)equipped with an evaporative light-scattering detector (ELSD, Waters 2420, Waters Analytical Instrument Co., Ltd., USA) and a Lichrospher C18 HPLC column (Hanbon Sci&Tech, Size: 4.6×250mm) . Methanol was used as the eluant at a flow rate of 0.8 mL/min.

The product bis (α-carboxyl-α-decyl) tetramethylammonium was characterized by LC-ESI-Q/ToF MS[Waters MALDI SYNAPT Q-ToF MS, Waters ACQUITY Ultra Performance Liquid Chromatograph (UPLC) & Xevo (TM)TQ] .

Results and discussion

A main by-product α-dimethylaminolauric acid may exist sometimes in the raw product, its structure as following:

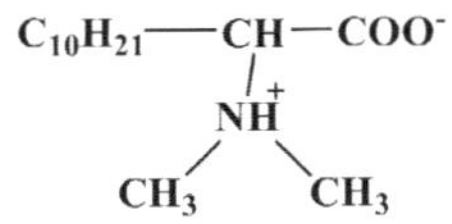

α-dimethylaminolauric acid

HPLC-ELSD affords a means of detecting impurities. α-dimethylaminolauric acid (by-product, abbreviated as α-DMALA) and α-chlorododecanoic acid (abbreviated as α-CDA) may be considered as the most probable impurities in bis (α-carboxyl-α-decyl) tetramethylammonium (abbreviated as BCDTA). The results of HPLC-ELSD for product show, that these impurities can be detected by this means. The refined product contains little these impurities. BCDTA, α-CDA and α-DMALA can be separated and detected clearly by HPLC-ELSD (Figure 1).

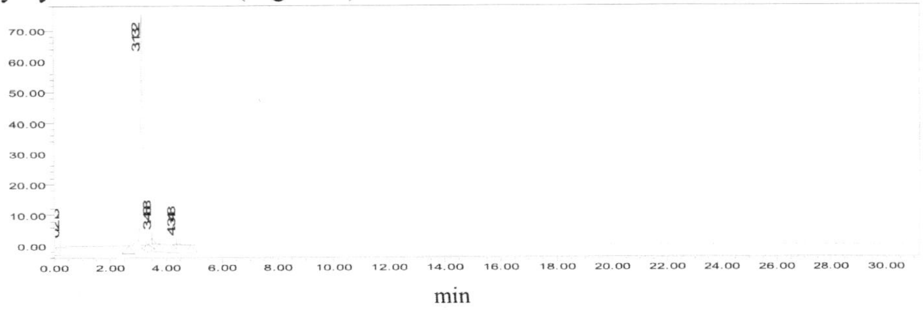

Figure 1 A) high performance liquid chromatograph of BCDTA

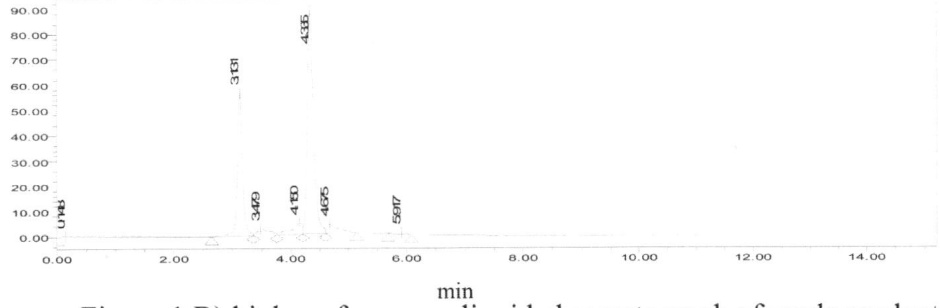

Figure 1 B) high performance liquid chromatograph of crude product

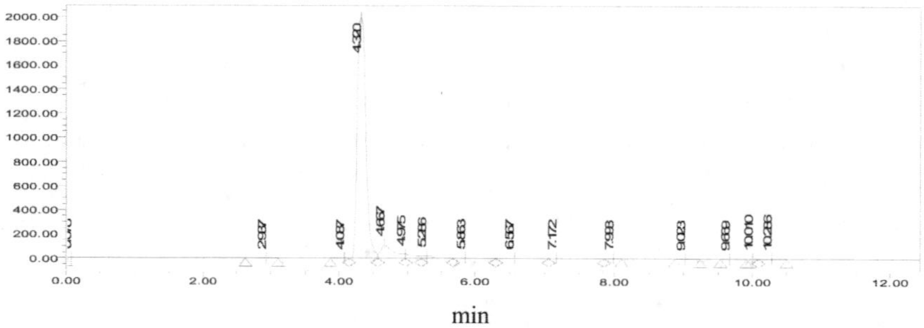

Figure 1 C) high performance liquid chromatogram of α-DMALA

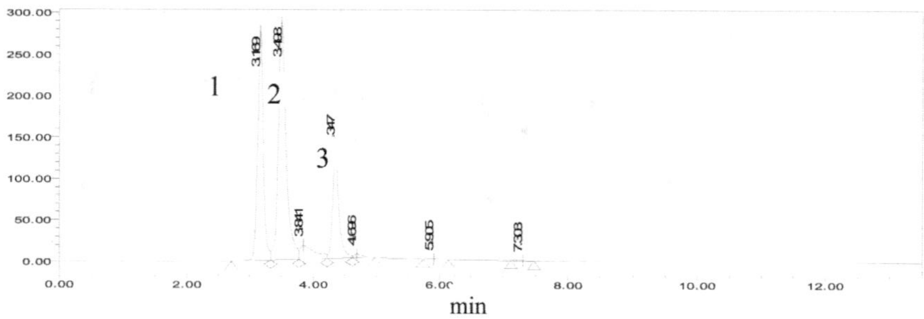

Figure 1 D) high performance liquid chromatogram of BCDTA (1) + CDA(2) + α-DMALA(3)

The molecular structure of the synthesized product was characterized by Mass spectrum (Figure 2), and was confirmed to be in good agreement with that of the target chemical.

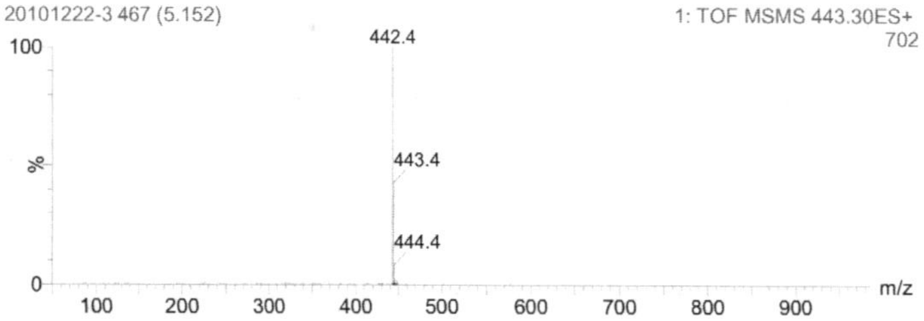

Figure 2 Mass spectrum (positive ion mode) of BCDTA

ESI-MS in the positive ion mode gives only peaks of pseudo molecular ions of the form $[M+H]^+$ which mainly correspond to protonated molecular ions with a small amount of fragments. Thus, the identity of the target product can be deduced by accurate determination of the molecular weight.

Conclusions

The novel Gemini betaine surfactants bis (α-carboxyl-α-decyl) tetramethylammonium (BCDTA) was synthesized by a two steps pathway from biomass raw material natural fatty acids. The purity of the refined product was confirmed by HPLC-ELSD, and was characterized by LC-ESI-Q/ToF MS. The molecular structure of the synthesized product was confirmed to be in good agreement with that of the target chemical.

References

1. Xie Z., Feng Y. (2010), "Synthesis and Properties of Alkylbetaine Zwitterionic Gemini Surfactants", *Journal of Surfactants and Detergents*, January, pp. 51–57.
2. Zana R. (2002), "Dimeric and Oligomeric Surfactants. Behavior at Interfaces and in Aqueous Solution: a Review", *Advances in Colloid and Interface Science*, March, pp. 205–253.
3. Williams A.C., Barry B.W. (2004), "Penetration Enhancers", *Advanced Drug Delivery Reviews*, March, pp. 603–618.
4. Chevalier Y. (2002), "New Surfactants: New Chemical Functions and Molecular Architectures", *Current Opinion in Colloid & Interface Science*, March, pp. 3–11.
5. Candau S.J., Oda R. (2001), "Linear Viscoelasticity of Salt-free Wormlike Micellar Solutions", *Colloids and Surfaces A: Physicochemical and Engineering Aspects,* July, pp. 5–14.
6. El Achouri M., Infante M.R., Izquierdo F., Kertit S., Gouttaya H.M., Nciri B. (2001), "Synthesis of Some Cationic Gemini Surfactants and Their Inhibitive Effect on Iron Corrosion in Hydrochloric Acid Medium", *Corrosion Science*, January, pp. 19–35.
7. Laschewsky A., Wattebled L., Arotcarena M., Habib-Jiwan J.L., Rakotoaly R.H. (2005), "Synthesis and Properties of Cationic Oligomeric Surfactants", *Langmuir* , July, pp. 7170–7179.
8. Sharma K.S., Rodgers C., Palepu R.M., Rakshit A.K. (2003), "Studies of Mixed Surfactant Solutions of Cationic Dimeric (gemini) Surfactant with Nonionic Surfactant $C_{12}E_6$ in Aqueous Medium", *Journal of Colloid and Interface Science,* December, pp. 482–488.
9. Yoshimura T., Esumi K. (2004), "Synthesis and Surface Properties of Anionic Gemini Surfactants with Amide Groups", *Journal of Colloid and Interface Science,* August, pp. 231–238.
10. Peresypkin A.V., Menger F.M. (1999), "Zwitterionic Geminis. Coacervate Formation from a Single Organic Compound", *Organic Letters, October*, pp. 1347–1350.
11. Alami E.O., Holmberg K. (2003), "Heterogemini Surfactants", *Advances in Colloid and Interface Science*, February, pp. 13–46.
12. Yoshimura T., Nyuta K., Esumi K. (2005), "Zwitterionic Heterogemini Surfactants Containing Ammonium and Carboxylate Headgroups. 1. Adsorption and Micellization", *Langmuir*, February, pp. 2682–2688.
13. Yoshimura T., Ichinokawa T., Kaji M., Esumi K. (2006), "Synthesis and Surface-active Properties of Sulfobetaine-type Zwitterionic Gemini Surfactants", *Colloids and Surfaces A: Physicochemical and Engineering Aspects*, February, pp. 208–212.
14. Paatero E., Salmi T., Fagerstolt K. (1992), "Selective Synthesis of α-Chlorocarboxylic Acids", *Industrial & Engineering Chemistry Research,* November, pp. 2425-2437.
15. Qi L., Fang Y., Wang Z., Ma N., Jiang L., Wang Y. (2008), "Synthesis and physicochemical investigation of long alkylchain betaine zwitterionic surfactant", *Journal of Surfactants and Detergents*, February, pp. 55–59.
16. Wang Z., Zhang S., Fang Y., Yang H., Qi L., (2010), "A Kinetics and Mechanism Investigation of the Nucleophilic Substitution Reaction of a-Chlorododecyl Carboxylate with Trimethylamine", *Journal of Surfactants and Detergents*, January, pp. 361–365.

Solar absorption cooling with alkanes as phase change energy storage medium

L. Agyarko and G.A. Mansoori
Department of BioEngineering
University of Illinois at Chicago

Introduction

In this presentation we report our research in the development of a cooling system based on the photothermal solar energy utilization, phase change energy storage in alkanes and a low temperature gradient absorption cooling cycle, the areas in which our research in the past has resulted with well-defined characterization methods. The performance parameters considered in our work are collector surface requirement [1], coefficient of performance (COP) [2, 3, 5], capacity and size of the storage medium for the cloudy days and evenings [4], heat duties of the evaporator, absorber, and condenser, refrigerant nature and concentration in the weak and strong solution [2, 3, 5], and flow rates of the weak solution and the flow rate of refrigerant passing through the evaporator.

The initial cost of investment for solar absorption cooling systems is very high, compared to compression-expansion cooling systems. To offset the high initial cost, a combined air conditioning system alternating between cooling and heating, using the same basic equipment and collector is often proposed. In recent years, the increasing high cost of fusil fuels, coupled with its adverse environmental impact has aroused much interest in solar cooling technology. For these reasons, while solar energy technology is already used for heating, its potential for cooling is now quite appropriate to be exploited.

Solar energy technology can be used for cooling using Rankine cycle, absorption cycle [1] and desiccant cooling cycle. The principles of the absorption cycle method was earlier studied by Faraday in 1824 and was subsequently improved to provide cooling to localities devoid of electricity, using liquefied petroleum gas (LPG) or natural gas as heat source [1]. Currently, the two most commonly used absorption air-conditioning systems on the market are the single effect and double effect chillers. Both chillers use water-lithium bromide as refrigerant-absorbent pair and are favorable for solar absorption cooling. A significant step, toward the improvement of the efficiency of solar absorption cooling systems is the integration of thermal energy storage unit to balance the energy demand during cloudy days and nights. Several phase change materials like water, alkanes and paraffin wax and ionic liquids have been identified for low, intermediate and high energy temperatures for applications, such as latent thermal energy storage media [4]. However, the search for working fluids combinations for optimum design of solar absorption cooling systems remains a major challenge.

Process description of continuous solar absorption cooling system

The design of the continuous solar absorption cooling system is shown in figure1 [5]. Solar radiation from the sun is transformed into heat by the solar collector (flat-plate). The heat is stored in the phase change energy storage medium (alkane) and is transferred to the generator to vaporize the refrigerant from the absorbent, using hot water from the water tank as heat transfer medium. The vaporized refrigerant then enters the condenser (high pressure side), where it is liquefied through heat removal using air or cooling water at room temperature. The condensed refrigerant is again vaporized through an expansion valve to the evaporator (low pressure side) to produce cooling. The vaporized refrigerant then enters the absorber, dissolving in the weak refrigerant-solvent solution to form a strong refrigerant solution. Since the mixing of the refrigerant with the solvent in the absorber is exothermic process, heat is removed by air or cooling water at room temperature. This heat removal creates a strong chemical affinity between the refrigerant and the weak solution. The rich solution formed is recycled to the generator through the pump to repeat the process [3].

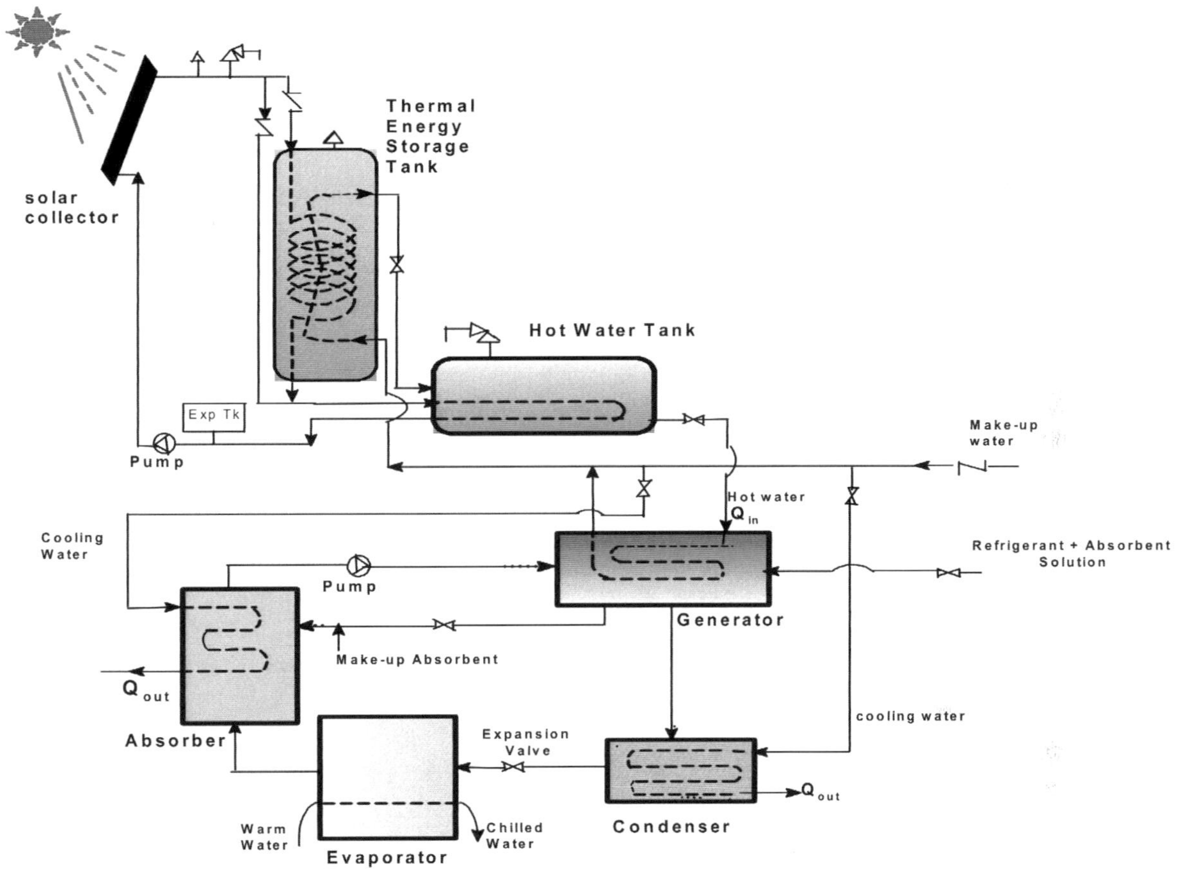

Fig. 1 Design of continuous solar absorption cooling system with thermal energy storage unit [5]

Solar Collector Plate Performance

The energy collection efficiency (η_{col}) of a flat-plate collector is usually defined as [5]

$$\eta_{col} = F_R\eta_{opt} - F_R U_l \left(\frac{T_{in} - T_a}{I_{t,a}}\right) \quad (1)$$

Where

η_{opt} = optical efficiency
U_l = overall heat loss
$I_{t,a}$ = total aperture irradiance
F_R = collector heat removal efficiency factor
T_{in} and T_a are the inlet and outlet temperatures of the heat transfer fluid respectively

The equation for calculating solar collector area (A_a) is defined as

$$A_a = \left(\frac{\dot{Q}_{useful}}{\eta_{col} . I_o}\right) \qquad (2)$$

Where

$\dot{Q}_{useful}$ = rate of (useful) energy output (KW)

A_a = aperture area of the collector (m^2)

I_O = solar irradiance falling on collector aperture (W/m^2)

Storage of latent thermal energy using paraffin waxes

The storage of latent thermal energy in a phase change material (PCM) such as alkanes and paraffin waxes is achieved by a phase transformation in the medium. Two phenomena occur during the phase change: melting and freezing. During the melting process, latent heat is absorbed by the medium, while in freezing heat is released. In phase transformation the temperature of the medium remains constant, while latent thermal heat is being stored. In essence, the PCM utilizes some sensible heat storage, below and above its melting point temperature which is defined as [4]

$$Q = m\left[\int_{T_1}^{Tm} C_{ps}\, dT + L + \int_{T_m}^{T2} C_{pl}\, dT\right] \qquad (3)$$

Where

m is the mass of the PCM

L is its latent heat of melting

T_1, T_2 and T_m are the initial, final and melting absolute temperatures

C_{ps} and C_{pl} are specific heats of solid and liquid of PCM phases respectively

The first term of the equation represents the sensible heat of solid PCM phase (below melting point), while the second term represents the PCM latent heat of melting

Assuming that the alkanes present in the paraffin wax forms an ideal solution, then the total heat stored in the paraffin wax will be [4]

$$Q = \sum_i m_i\left\{\int_{T1}^{T_{mi}} C_{psi}\, dT + Li + \int_{T_{mi}}^{T2} C_{pl}\, dT\right\} \qquad (4)$$

In equation (4), the summation is over the components of the paraffin wax and every term with subscript i is a property of the ith component (alkanes) of the paraffin wax. The cumulative thermal energy which is an important thermal energy storage characteristic of paraffin wax is defined as the thermal energy stored in a wax by raising its temperature from a reference temperature (298 K in the present case) to a higher temperature, taking into account the composition effects of all the hydrocarbons present in the wax[4].

Thermodynamic requirements for the choice of working fluids

The major thermodynamic parameters affecting the performance of absorption solar cooling systems are (i) the solar collector efficiency, (ii) the coefficient of performance of the cycle or $(COP)_{cycle}$,(iii) the ratio $(\frac{\dot{m}_t}{\dot{m}_p})$, which is the ratio of the mass flow rate of the refrigerant throttled per unit mass of solution pumped from the low pressure to high pressure side, through the liquid pump and (iv) the pumping work of the strong solution from low pressure in the absorber to the high pressure side in the generator per ton of refrigeration, which is defined as[2]

$$\frac{|W_p|}{|Q_E|} = \frac{\dot{M}p}{\dot{M}t} . \frac{1}{\rho_A} . \frac{(P_G - P_A)}{(h_4 - h_3)} \qquad (5)$$

Where

ρ_A = density of the strong solution pumped from the absorber to the generator

P_A = absorber pressure

P_G = generator pressure

and h_3 and h_4 are the inlet and outlet enthalpies of the refrigerant passing through the evaporator

The parameter $\frac{|\dot{W}p|}{|\dot{Q}_E|}$ is the measure of pumping energy consumed per unit of cooling energy effect. The parameter $\frac{\dot{M}p}{\dot{M}t}$ is the measure of pumping required per unit refrigerant circulated through the cooling circle. For optimum design of absorption cooling cycles, smaller values of these ratios are required. In the comparative study of working fluid combinations, the COP, the upper bound of $\frac{\dot{M}t}{\dot{M}p}$ and $\frac{|\dot{W}p|}{|\dot{Q}_E|}$ are used as the criteria.

Thermodynamic analysis of solar absorption cooling system

From thermodynamic analysis, the theoretical coefficient of performance of an absorption cooling cycle is the ratio of the cooling effect by the evaporator and the heat input to the generator defined as [2]

$$(\mathrm{COP})_{cycle} = \frac{|Q_E|}{|Q_G|} \qquad (6)$$

Where

Q_E is cooling effect by the evaporator

and Q_G is the heat input to the generator

Similarly, the overall coefficient of performance of solar absorption cooling system is the ratio of the cooling effect by the evaporator and the solar energy input to the collector defined as [2]

$$(\mathrm{COP})system = \frac{|Q_E|}{|Q_S|} \qquad (7)$$

Where

Q_E is cooling effect by the evaporator

and Q_S is the solar energy input to the collector

Dividing equation (7) by equation (6) yields the efficiency of the solar collector defined as [2]

$$\eta collector = \frac{|Q_G|}{|Q_S|} \qquad (8)$$

Where

Q_G is the heat input to the generator

and Q_S is the solar energy input to the collector

Consequently, the overall coefficient of performance of the solar absorption cooling system is [2]

$$(\mathrm{COP})system = (\mathrm{COP})_{cycle} \cdot \eta collector \qquad (9)$$

According to the first law of thermodynamics, the following equation exists for the absorption cycle [2]

$$Q_G + Q_E - Q_C - Q_A + W_P = 0 \qquad (10)$$

Also, according to the second law of thermodynamics the following relation can be written for the absorption cycle [2]:

$$\frac{\Sigma |Q_i|}{T_i} \leq 0$$

or

$$\frac{|Q_G|}{T_S} + \frac{|Q_E|}{T} - \frac{|Q_C|}{T_o} - \frac{|Q_A|}{T_o} \leq 0 \quad (11)$$

By assuming W_p, the work input to the liquid pump negligible, as compared to the other terms in equation (10) and joining the resulting relation with inequality (11) we get [2]

$$\frac{|Q_G|}{T_S} + \frac{|Q_E|}{T} - \frac{1}{T_o}(|Q_C| + |Q_A|) \leq 0 \quad (12)$$

now by consideration of the definition of $(COP)_{cycle}$ as given by equation (6), the above inequality can be arranged in the following form [2]

$$(COP)_{cycle} = \frac{|Q_E|}{|Q_G|} \leq \frac{T}{T_S}\left(\frac{T_S - T_o}{T_{G-T}}\right) \quad (13)$$

This inequality for COP can also be derived through the definition of Carnot cycle and its efficiency. The equality sign in equation (13) holds for the COP of a Carnot cycle. The right hand side of the inequality gives the maximum COP which may be achieved for an absorption cooling cycle with known temperatures for environments of the components of the cycle. The maximum COP can be achieved only when the cycle operates reversibly [2].

In order to relate the characteristics of the cycle to the properties of the working fluids (refrigerant and absorbent), the laws of thermodynamics for flow systems (open-systems) can be utilized. According to the first law of thermodynamics for flow systems, the following relations exist between the heat and work transfer rates and the properties of the working fluids in a steady state absorption cycle [2]:

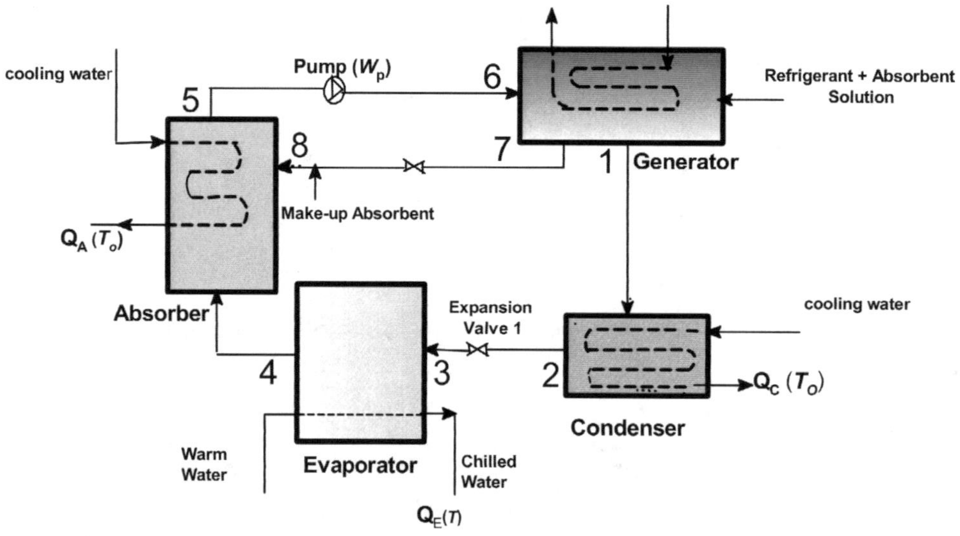

Fig 2 Schematic diagram of continuous solar absorption cooling cycle

$$|\dot{Q}_G| = \dot{M}_t(h_1 - h_7) + \dot{M}_P(h_7 - h_6) \quad (14)$$

$$|\dot{Q}_C| = \dot{M}_t(h_1 - h_2) \quad (15)$$

$$|\dot{Q}_E| = \dot{M}_t(h_4 - h_3) \quad (16)$$

$$|\dot{Q}_A| = \dot{M}_t(h_4 - h_8) + \dot{M}_P(h_8 - h_5) \quad (17)$$

$$|\dot{W}_P| = \dot{M}_P(h_6 - h_5) \quad (18)$$

In the above equation, h_1 is the specific enthalpy of point 1 in figure 2, M_t is the mass flow rate of refrigerant passing through the throttling valve and M_P is the mass flow rate of the strong solution passing through the liquid pump. Also, the following relations exist between M_t and M_P [2]

$$\frac{\dot{M}_t}{\dot{M}_P} = (X_A - X_G)/(1 - X_G) \qquad (19)$$

In equation (19) X_A is the mass fraction of the refrigerant in the strong liquid phase coming out of the absorber and X_G is for the liquid phase coming out of the generator. In deriving this equation, it is assumed that the vapor which is coming out of the generator is the pure refrigerant. The following relations also exist for the isenthalpic expansion valves in the cycle [2]

$$h_2 = h_3 \text{ and } h_7 = h_8 \qquad (20)$$

By assuming W_p, the power input to the liquid pump negligible, the following relations will also be derived [2]

$$h_6 = h_5 \qquad (21)$$

Based on the above equations, the COP of the cycle can be expressed with respect to the working fluid properties at different locations of the cycle using the following equation [2]

$$(\text{COP})\text{cycle} = \frac{|\dot{Q}_E|}{|\dot{Q}_G|} = \frac{h_4 - h_2}{\left[(h_1 - h_7) + \frac{(1-X_G)(h_7 - h_5)}{(X_A - X_G)}\right]} \qquad (22)$$

Where

$\dot{Q}_E$ is cooling effect by the evaporator

$\dot{Q}_G$ is the heat input to the generator

X_A and X_G are mass fractions of the refrigerant in the solutions leaving the absorber and generator respectively and h_1, h_2, h_4 and h_7 are the molar enthalpies of the refrigerant at the specified locations of the absorption cooling cycle (fig 2).

According to the second law of thermodynamics for open systems, the following relations also exist between the heat transfer rates and the properties of the working fluids in a steady-state absorption cycle [1]:

$$|\dot{Q}_G| \le T_S\{\dot{M}_t(s_1 - s_7) + \dot{M}_P(s_7 - s_6)\} \qquad (23)$$

$$|\dot{Q}_C| \ge T_O\dot{M}_t(s_1 - s_2) \qquad (24)$$

$$|\dot{Q}_E| \le T\dot{M}_t(s_4 - s_3) \qquad (25)$$

$$|\dot{Q}_A| \ge T_O\{\dot{M}_t(s_4 - s_8) + \dot{M}_P(s_8 - s_5)\} \qquad (26)$$

In the above relation, the equality sign is for the reversible case and the inequality sign is for the irreversible case. By joining equations (23) and (25), we get [1]

$$|\dot{Q}_G| + |\dot{Q}_E| \le \dot{M}_t\{T_S(s_1 - s_7) + T(s_4 - s_3)\} + \dot{M}_P T_S(s_7 - s_6) \qquad (27)$$

Also by joining equations (24) and (26) we get [2]

$$|\dot{Q}_C| + |\dot{Q}_A| \ge \dot{M}_t\big(T_O(s_1 - s_2) + T_O(s_4 - s_8)\big) + \dot{M}_P T_O(s_8 - s_5) \qquad (28)$$

Again, by assuming that W_P is negligible as compared to the other terms in equation (10) we can write [2]

$$|\dot{Q}_G| + |\dot{Q}_E| = |\dot{Q}_C| + |\dot{Q}_A| \qquad (29)$$

From equations (27)-(29) we conclude that [1]

$$\dot{M}_t\{T_o(s_1 - s_2) + T_o(s_4 - s_8)\} + \dot{M}_P T_o(s_8 - s_5) \le |\dot{Q}_G| + |\dot{Q}_E| \le \dot{M}_t\{T_S(s_1 - s_7) + T(s_4 - s_3)\} + \dot{M}_P T_S(s_7 - s_6) \qquad (30)$$

By dividing relation (30) by $|\dot{Q}_G|$ and considering the definition of $(COP)_{cycle}$ as given by equation (6) we get the following relation [2]

$$LL \leq (COP)cycle \leq UL \tag{31}$$

Where the lower limit (*LL*) is in the following form [2]

$$LL = \frac{\left(\frac{\dot{M}_t}{\dot{M}_P}\right)\{T_o(S_1-S_2)+T_o(S_4-S_8)\}+T_o(S_8-S_5)}{\left(\frac{\dot{M}t}{\dot{M}p}\right)(h1-h7)+(h7-h5)} - 1 \tag{32}$$

and the upper limit (*UL)* is the following form [2]

$$UL = \frac{\left(\frac{\dot{M}_t}{\dot{M}_p}\right)\{T_s(S_1-S_7)+T(S_4-S_3)\}+T_s(S_7-S_6)}{\left(\frac{\dot{M}t}{\dot{M}p}\right)(h_1-h_7)+(h_7-h_5)} - 1 \tag{33}$$

Where

M_t = mass flow rate of the refrigerant going through the throttle valve to the evaporator

M_p = mass flow rate of the strong solution pumped from the absorber to the generator

h_1, h_5, h_7, S_1, S_2, S_3, S_4, S_5, S_6, S_7 and S_8 are the molar enthalpies and entropies of the refrigerant at the specified locations of the absorption cooling cycle (fig 4)

T_S = hot source temperature added to the generator

T_O = environmental temperature

T = evaporator temperature

Relations (31)-(33) can be used to calculate the upper and lower limits of the cycle COP at different working conditions. This means that the cycle COP will remain in the bounds as specified by equation (31). However, it should be pointed out that the actual COP of the cycle is even less than the COP as calculated by equation (22). Also with the understanding that inequality (13) gives the upper limit of COP, regardless of the working fluids under consideration, and since it is for a Carnot cycle, it is always higher than *UL* as given by equation (32). Consequently, we can write [2]

$$LL \leq (COP)_{actual} \leq (COP)_{cycle} \leq UL \leq \frac{T}{T_s}\left(\frac{T_s-T_o}{T_o-T}\right) \tag{34}$$

For evaluation and comparison of different working fluids, equation (34) can be used. But prior to the evaluation and comparison, a number of preliminary criteria should be specified. For example, the criteria for selecting those working fluid combinations which possess the basic and necessary physiochemical and environmental characteristics as refrigeration working fluids should be specified.

Results and Discussion

Figure 3 shows the profile of the cycle COP and its upper and lower bounds for NH_3-H_2O refrigeration cycle for different solar collector temperatures using experimental data. The expected or cycle COP was calculated using equation (22), and similarly, its lower and upper bounds were calculated from equations (32) and (33) respectively. The temperature of the solar collector (Ts) ranged from 85°C to 100°C at two different environmental temperatures (T_o = 21.1°C and T_o = 26.7°C) and evaporator temperature (T_E = 2.2°C). The shaded areas represent the cycle COP operating regions for the two different environmental temperatures. However, the actual COP operating region as represented by the inequality (34) is less than the cycle COP operating area confined between the cycle COP and its lower bound, without reaching the upper bound. Comparing the cycle COP for the two environmental temperatures (T_o = 21.1°C and T_o = 26.7°C), it is clear that for refrigeration systems, the cycle COP for NH_3-H_2O cycle is less sensitive to changes in solar collector temperature. However, high sensitivity of cycle COP of refrigerant-absorbent pairs to solar collector temperature is one of the criteria required when screening refrigerants [2, 3]. For comparative study of refrigerant-absorbent pairs for solar absorption cooling systems, the two other criteria mentioned earlier in this paper will be required.

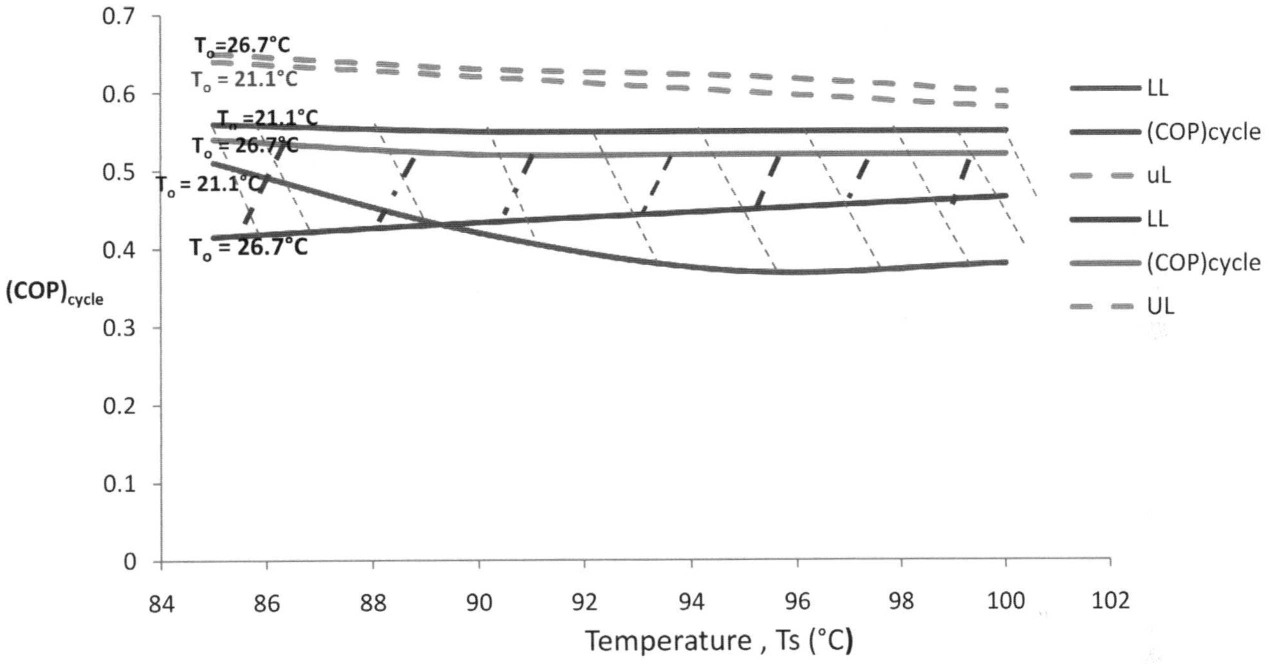

Figure 3 $(COP)_{cycle}$ profile of NH_3-H_2O cycle and its upper and lower limits versus solar collector temperature (Ts), for a refrigeration cycle with evaporator temperature (T_E =2.2°C) and environmental temperatures (T_o= 21.1 °C and T_o = 26.7°C).

References:

[1]. G.A. Mansoori, "Prospects for Solar Energy Utilization - Photothermal Methods", IJST, Vol.5, No.2, pp.55-73, 1976.

[2]. G.A. Mansoori and V. Patel, "Thermodynamic Basis for the Choice of Working Fluids for Solar Absorption Cooling Cycles", Solar Energy Journal, Vol.22, pp.483-491, 1979.

[3]. A.L. Gomez and G.A. Mansoori, "Thermodynamic Equation of State Approach for the Choice of Working Fluids of Absorption Cooling Cycles", Solar Energy Journal, Vol. 31, pp.557-566, 1983.

[4]. S. Himran, A. Suwono, and G.A. Mansoori, "Characterization of alkanes and paraffin waxes for application as phase change energy storage medium", Energy Sources, Vol. 16, pp. 117-128, 1994.

[5]. L. Agyarko and G.A. Mansoori, "Cooling and Air Conditioning through Solar Energy", Proceedings of the Alternative Energy Symposium 2010, Chicago, IL Oct. 7, 2010.

Role of University Startup Companies in Technology Commercialization

Robert F. Anderson
Director Technology Transfer and Intellectual Property
Illinois Institute of Technology

For presentation at the AIChE Spring Meeting
13-18 March 2011, Chicago, IL USA

Abstract

The traditional university role of performing fundamental research leaves a large gap between the laboratory stage "proof of principle" and the generation of economic benefits in the commercial marketplace. Research funding for universities from the federal government tends to focus on basic research. Companies usually prefer to license technology with most of the risk taken out. This risk reduction normally occurs via the development component of the research and development process. Unfortunately, the cost of D is often 10 plus times the cost of R, which puts this effort out of reach of most universities. In addition, the opportunity for publication and contribution to advancing the state of the art science (the usual focus of university research) is limited as the development activity concentrates on such issues as economic tradeoffs, catalyst life, feedstock impurities, manufacturing processes, corrosion and other materials of construction issues, product specifications, and mechanical hardware problems.

To bridge this gap, Illinois Institute of Technology has put in place a series of practices to address the problems that stand in the way of the university's ability to pursue commercial success for new technology. Our primary practice is to establish startup companies by the inventing faculty members who continue to guide the technology development process into the commercial marketplace. Additional contributors are undergraduate students working on these more-practical problems, mutually beneficial strategic alliances with existing commercial companies, incubator facilities with their ancillary networking potential, senior executives and alumni acting as friends of the university in mentoring entrepreneurial faculty, the technology transfer function of the university, the university's marketing and communications professionals, and alignment of startup company interests with the university's goals via equity investment in the company. This presentation describes how these practices have also provided benefits to our development partners from industry

Introduction

Political leaders, business people, and academics have identified problems with US competitiveness, especially in the manufacturing sector. Yet the gap in technology interaction that still exists between university and industry seriously diminishes the strength of the US business community relative to its peers around the world. This is manifested by students and graduates with little understanding of the real world, companies that lack a window to technological developments in distant countries, companies that lack first hand knowledge of foreign cultures, and academics working on obscure problems rather than issues of pressing interest to the industrial community.

I do not mean to imply that any one segment is responsible for failing to achieve the potential synergy. Rather, I indict the lack of, or weakness of, processes to ensure rapid identification of commercially significant problems and equally rapid generation of solutions. In my industrial career I had the privilege of managing a department of seasoned professionals whom I defined as bilingual in business and technology. Employees selected for this group had to have the ability to communicate with basic researchers about problem definition and about the equally difficult solution recognition. Problem definition had to start with marketplace needs that they translated into research goals – a new product, modified product specifications, a new raw material, a new processing route, a significant cost reduction, or whatever. Solution recognition might involve a catalyst formulation with more selectivity, stability, or greater activity; an innovative reactor design; new approaches to product separation or purification; or clever chemistry. The next step from research to development represented a challenge much like we in technology transfer face when we evaluate results from the university laboratories.

The accomplishments during the development stage may seem rather modest to some, but without them the idea remains a note in the researcher's logbook or perhaps in a patent that languishes with no commercial use. Pilot plants; computer models; physical or mechanical models; scale-up techniques; metallurgy and other materials of construction; sensors and analytical techniques; feedstock evaluation, selection, and preparation; long-term demonstration of catalyst life and process stability; catalyst manufacture and regeneration; hardware design and development; physical property measurement; hazard identification and mitigation; environmental issue identification and mitigation; and process engineering – the list can seem endless. The development process can take an order of magnitude more time than the basic research and require an order of magnitude more money. We believe we have identified many of the barriers and demonstrated novel ways to get around them as developed below.

Industry/University Interface: A Necessary Bridge

Barriers to technology commercialization involving industry/university collaboration include differences in several important areas:

- Time frames (semesters vs. days)
- NIH (Not Invented Here)
- Expectations (Good enough vs. perfectly understood)
- Resource constraints
- Reward systems
- Culture
- Profit motive
- Intellectual Property issues
- Educational mission
- Publication policy

New models of university technology transfer have the potential to address many of the challenges of R&D process effectiveness in academia but also the adoption of new technology by the industrial sector. However, the university side needs to recognize that challenging and interesting problems abound in the industrial sector and industry people need to recognize that most professors do not want to sit in an ivory tower and contemplate arcane theories.

Role of University Startup Company

To cut the time required for invention, development, demonstration, and commercialization of new ideas from university research labs requires addressing the cultural, systems, process, and organizational issues listed above. Credibility for the new technology can be enhanced by launching a startup company that gets product into the marketplace, solves production problems, defines the market opportunity, and then generally dissolves through a merger or acquisition process with a mature company in the industry. However, in the first sentences of this paper, I identified the traditional university's lack of interest in product and process development as the root cause of much of the failure to commercialize technology. Absent a support structure that provides business expertise to a budding entrepreneur, even the most motivated professor will give up in the face of the many barriers a new company faces. The problem to be solved is thus "how do we make that support available given the limited resources of the typical academic institution?"

Approach

Four critical topics must be addressed:

- Attracting experienced management personnel
- Funding the early stage venture
- Accessing university intellectual property
- Changing focus from technology to identifying and satisfying customer needs

The first step is to make the inventor aware that being a company president requires a skill set very different from their academic position. They can choose to take a crash course, learn by trial and error, or seek professional help. In reality, they will need to exercise all three options. An active board of volunteer advisors can be helpful in avoiding major mistakes. Creative use of equity-based compensation can supplement cash salaries that fall short of what is normally required to attract executive talent. Next is to develop an appreciation for the fundraising process that will consume most of the entrepreneur's waking hours for the next ten years. FFF (friends, family, and fools), angel groups, alliance partners, venture capitalists, SBIR grants, and other government programs are potential sources of investment capital, but all involve significant time and effort to keep up with geometrically expanding money needs.

Intellectual property issues must be solved early. Most university policies require assignment of IP, just as most company policies require. We have decided that the value added by solving the commercialization issues justifies offering an exclusive license to the essential technology in return for a modest equity stake (~10%). Recognition that the university stands to gain from success of the business can help support policies that align with the needs of the entrepreneur.

The transition from focus on technology to focus on customers, competitors, markets, business models, and distribution options is difficult for a technologist. Unfortunately, it is failure to make this shift that brings most new ventures down. Owning a factory sounds like fun to many engineers, but reality usually intervenes in favor of some version of outsourcing.

Any new product faces resistance in the marketplace. The usual excuses of "current product is good enough, I'd have to change other aspects of my product, too risky, happy with current suppliers" are supplemented by "never heard of you, how do I know you will be around next year, how would you handle a big order, etc.?"

Learning real customer needs is much harder than it sounds. Once this is achieved, demonstrations of prototypes are an effective door opener. They are also very expensive. This is another example of the value of a strategic partner who can bear some of the expense and open the right doors to potential purchasers.

Results

We have successfully transferred technology into the marketplace from research programs in chemical engineering and electrical engineering. A number of problems surfaced and have been solved to a significant degree. Lessons learned will enable faster commercialization of future inventions. Additional opportunities for improvement in the impact of the new technology have been identified and will be tested in the near future.

Case Study

Thermal Management of Lithium Ion Batteries (Chemical engineering)

A government research grant received at IIT fifteen years ago had the goal of generating understanding of the phenomena surrounding heat release from lithium ion batteries. The application envisioned by the funding agency was portable power for military troops. In recognition of the increased need for power in the field for computers, night vision goggles, cooling systems for personal comfort, radios, and other electronic equipment, lithium ion battery technology had attractive potential based on high energy to weight ratio, minimal self-discharge, fast charge capability, and long shelf life. Unfortunately, thermodynamics got in the way by creating concern for the potential of thermal runaway.

IIT professors J. Robert Selman and Said Al-Hallaj conducted a study of the fundamentals of heat generation from lithium batteries in which they identified a number of interesting phenomena. If this had been the usual academic research project, they would have published papers on energy, entropy, calorimetry, reaction kinetics, and the like and gone on to other activities. However, given the government's interest in getting safe, lightweight batteries into the field and the professors' motivation to see their research applied in the real world, the project continued with specific product goals.

First, the professors thoroughly defined the problems to be solved by their product design. The major problem was that the required total energy at the energy density specified creates a significant risk of thermal runaway, fire, and possibly even an explosion. Additional finished product requirements included:

- Light weight
- Shape capable of being customized to fit space constraints
- Not combustible
- Capable of removing a large amount of heat
- Maintain uniformity of temperature between individual cells in a large battery pack
- No individual cell temperature rises above a threshold limit that would trigger cell degradation or temperature runaway
- Technology applicable to different cell sizes and configurations,
- Compatible with future improvements in battery chemistry
- Inexpensive

This sounds like a tall order, but it provides a clear set of criteria for evaluating suggested solutions.

Like most elegant solutions, once identified they become obvious. The need for high heat absorption capacity suggests a phase change phenomenon. Melting seemed simpler than boiling. The desired maximum temperature of roughly 50° C suggested a hydrocarbon wax. Initial trials confirmed that a wax melting at 50° did an excellent job of taking up the heat and limiting maximum cell temperature to the wax melting point. Unfortunately, molten wax is also an excellent insulator, which interfered with conducting the heat away from the battery pack.

At this point, an undergraduate student team from IIT's IPRO™ (Interprofessional Projects) program was assigned the problem of heat conduction. They found an interesting expanded aluminum material that did a good job of carrying the heat away – once. Unfortunately, the wax pooled at the bottom of the container and was unavailable for the next cycle. It is fortunate that it did not work, as the cost of the aluminum alone was slightly higher than the proposed price for the finished product.

This failure created a clear goal for the next round of thermal conduction research. The material needed to be inexpensive, highly conductive, and needed the ability to hold the molten wax within its pore structure. At this point, experience with technology for an architectural application suggested that graphite might be the answer. A mechanical engineering student was dispatched to a university lab overseas to work with the faculty expert on the existing application and develop a program to transfer the technology to the battery application. He came back a year later with the problem solved.

The development plan suggested that early demonstration of a battery pack in a transport vehicle would be helpful in identifying additional technical problems. A different student team obtained a child's two wheel electric scooter, determined baseline performance, and substituted a lithium ion prototype. This began a rapid series of ever more challenging applications to bicycle, powered wheelchair, and neighborhood electric vehicle. Each of these extensions generated learning experiences disguised as failures. Motors burned out, electronic circuits fried, connectors failed, soldered joints came unsoldered, and data logging systems choked on the volume of information. The teams quickly diagnosed the problems, tried a solution, usually had to replace that as well, but quickly achieved desired performance and went on to the next level of complexity where they repeated the failure to success pattern. The benefits of rapid prototyping were again confirmed.

At this point the startup company had been formed by the inventing professors with some of the initial employees selected from the earlier undergraduate teams. Their first demonstration platform was a Ford Escape Hybrid, obtained on loan from the City of Chicago, with the goal of converting it to plug-in hybrid. By this time, design procedures for both battery packs and electronics had matured, thus reducing the need for the dramatic series of failures and successes. Operation of the vehicle on city streets for two years provided extensive data to support further efforts including a similar conversion of a Toyota Prius.

An interesting observation of the student teams reveals that one secret of their success was diversity of educational backgrounds (academic majors), place of birth, previous mechanical experience, mechanical aptitude, theoretical knowledge, and previous job experience. From the beginning, the startup company, All Cell Technologies, LLC, has been staffed by a diverse team of chemical, mechanical, and electrical engineers, computer science experts, and electrochemical engineers. Their small workspace allows for none of the isolation that narrowly constituted teams work under in large organizations. Problem solutions have generally required the combined expertise from two or three disciplines.

As the company has matured, its capability has expanded in areas of cell and battery characterization, modeling of heat transfer phenomena, optimization of materials and fabrication techniques, and most important applications engineering. Manufacturing of the product originally specified by the Army introduced issues of quality assurance, packaging, and cost reduction.

With product and manufacturing technology well in hand, business development and marketing became the focus of management attention. Small startup companies are not in the viewfinder of the auto industry, despite the reality that early stage companies have solved many fleet electrification problems. The market development path therefore duplicated the early development program – that is, from the bicycle market to scooters, motorcycles, neighborhood electric vehicles, and finally automobiles. Commercial introduction often involves cycles of a handful of prototypes, a hundred demo units to confirm ruggedness, reliability, and life representations, and finally normal commercial orders – time consuming, expensive, but necessary.

Lessons Learned

To counter the root causes of failure, we have identified the root causes of success:

- A technically challenging problem that provides ample opportunity for the university to carry out its primary missions of education and expansion of knowledge
- An opportunity to use the solution to that problem to address an issue of importance to society
- Professors who have an interest in seeing the results of their creativity adopted by the world outside the campus
- Professors who are willing to learn and devote significant personal time and resources to making the commercialization process succeed
- Professors who are open to new ideas from anywhere
- A university policy, climate, and culture that encourages entrepreneurship
- Willingness of the university to forego short-term financial gain by taking equity in the startup business as compensation for technology licenses
- An incubator facility that provides a venue for focusing on the company
- Undergraduate students with work experience from summer work, coop jobs, and internships
- A component of the undergraduate student educational curriculum (e.g. IPRO™ – Interprofessional Projects) and experience (internships) that includes opportunities to work with the startup companies
- Coaching in the process of preparing business plans and assistance in finding investors who will provide funding for early stage companies
- Assistance from university leaders (senior management and trustees) in meeting prospective industrial partners, professional service providers, and executive talent
- A combination of strategically focused experimentation seasoned with serendipity from student workers who don't know what won't work
- Strategic partners from industry, individuals and organizations, who are willing to provide in-kind contributions, contacts, advice, shared advertising, and access to trade shows in return for early access to the next big idea
- Government officials who provide support from publicity to demonstration
- Finally, university staff (often in the technology transfer office) with significant executive-level experience with technology commercialization in industry

Benefits for Strategic Partners

- Preview of new technology in the pipeline
- Head start on introduction of new products
- Opportunity to recruit staff from the development teams
- Due diligence for potential acquisition

Conclusion

We believe that application of the guidelines and suggestions presented in this paper provide an opportunity for universities to enter the development stage of technology commercialization. However, collaboration between established firms and university startups appears to be the best solution for taking advantage of new technology.

Beyond University Research – Development Steps to Commercialization

Bipin Vora, Consultant, R&D Advisor;

For presentation at the AIChE Spring Meeting
13-18 March 2011, Chicago, IL USA

KEY WORDS: Research, development, catalysis, commercialization.

ABSTRACT

A lot of good research work is being done at many universities, but only a few get to commercial realities. University researchers, academics often wonder why industry executives or entrepreneurs or venture capitalists are not running to them to sign off to the next stage. The barrier is very high, often not visualized or understood. This presentation outlines all necessary steps one must go through from research to development, scale-up and commercialization.

A project from research commercialization has to pass through not one but many hurdles. One of the first is to pass techno-economic analysis. The investment needed for further development and scale-up of the technology and the duration of the time it would take are if not always, often significantly underestimated at the research stage. The author will discuss this and other criteria in more detail with a few examples.

INTRODUCTION

Continuing research leading to new innovations is the force behind the advancement of science and technology. Research at many universities has played an important role in this regard. Research conducted by a private or a public corporation is usually limited to the products and/or processes of their business. They have a built in user of their research output. On the other hand, research conducted at a university or a national laboratory is often looking at fundamental science, or trying to find a solution for a broader environmental or a societal problem and must find a first customer for its application. Research and development leading to commercialization of a new technology is not a trivial matter. It is an expensive, time-consuming undertaking. The many steps in concept to commercialization can be grouped in four phases.

Phase 1 - Idea Generation to Establishing Basic Process Concept

Laboratory research to confirm proof of principle
Conceptual Process and Catalyst Identified
Some Techno-economic and Market Analysis done

Phase 2 - Development Phase

Pilot Plant Design, Safety and Hazard analysis, Construction and Start-up
Pilot Plant Operation for

- Definition of Feed Quality requirements
- Determination of Product Quality
- Determination of long-term Catalyst stability, regenerability, regeneration method
- Kinetic Data, rate determination, develop kinetic model
- Obtain key information for detail Engineering Design
- Identify analytical methods for process and product quality control

Frequently revisit techno-economic & market analysis to assure continued project viability

Phase -3 Demonstration Phase

Scale-up of pilot plant data for design of the demonstration unit
Construction, Safety Check and Hazard analysis, and Start-up of the demonstration unit
Reconciliation of pilot plant and demonstration plant results

Phase 4 - First Commercial Plant
Scale-up to design of first commercial unit
Location and capacity of the first unit decided
Construction and start-up of the commercial unit
Trouble shooting, if needed
Meet process and product specification guarantees

If the process involves a new catalyst, then all above steps are repeated in parallel for the catalyst development and manufacturing scale-up. Thus, overall process becomes very complex and lengthy. From concept to commercialization the development of new technologies may take anywhere from 1 to 20 years based on relative simplicity or complexity of its application. Most often the development is a modification or advancement of an existing idea or process. For an example, an improved heat exchanger providing more energy efficiency or lower fouling tendency can easily be put in service in parallel to existing old heat exchanger in a commercial plant. Similarly, an improved catalyst for an existing process that may provide more stable operation, or may result in a higher product yield, can be tested by loading it in a commercial plant. These are very low risk proposal as old equipment or the old catalyst can be put back in service if any problems arise. In such applications, the demonstration phase becomes the first commercial application phase, thus substantially reduces time to commercialization as well as the total R&D cost.

Periodically a new process may come about for the same product because of increasing product demand or an availability of different advantageously priced starting raw material. Until about mid 1950s only source for propylene was fluid catalytic cracking unit in a refinery. With increasing demand for ethylene for production of polyethylene thermal steam cracking of naphtha and other hydrocarbons was developed that also provided propylene as a byproduct. These two sources for propylene could not keep up with increasing propylene demand that came about in mid 1980s and a selective catalytic dehydrogenation of propane to propylene was developed and commercialized in 1990. More recently a fourth route, methanol to propylene and ethylene is developed.

An introduction of a new product that requires both process development for the manufacturing of the product followed by product application is significantly more complex. These types of research and development require substantial time and expenditures as it would require all four phases of development work for the process as well as the product application. Planning and execution of any R&D program is critical to its success. Depending on the complexity of the process, demonstration units may easily cost $10 to 100 million U.S. or more. It requires a structured process; employing modern tools such as, Stage Gate Reviews, Failure Mode Effect Analysis (FMEA), etc. Technology development is not a straightforward process and failures abound – focus, resolve, and the appreciation of failures eventually lead to success. If we segment time and cost of these four phases of development one would see an *S* shaped curve.

The idea generation and proof of principle may take a long time, but is relatively less expensive, because of simplicity of laboratory equipment and for looking at limited information. Assuming that the results of first phase meet techno-economic criteria, the next phase of process development often moves fast. However, because of construction of a pilot plant closely resembling the commercial operation in a miniature scale, its operation requiring to collect lot of information that will be needed for a design and scale up to the demonstration and commercial units, it is a costly undertaking.

. .

Phase 1 - IDEA GENERATION

All new developments start from a seed idea. Necessity is the mother of invention. Often the new technology comes about because of a customer need, or an environmental change, or the advances in one technology area leading to advances in another. Only very few times does a truly creative innovation come about. **Technology inventions often are not identifiable with a single source or inventor but may be the result of a team effort.** Whether by design or by luck, inventions that originate at the laboratory level are more likely to be the inspiration of a single individual. Engineering inventions are quite often the product of a team, but a team requires a leader. He or she may be the originator of the idea or a colleague who sees the potential of such an idea – with a strong personality to nurture a seed idea to fruition and to lead the team effort that will eventually result in commercial implementation.

Research at universities is often limited to the first phase; idea generation, proof of principal, identification of catalyst and conceptual process flow scheme. There have been few rare cases where university research has gone all the way to a reasonable size demonstration unit. Often research at university becomes stagnated because its application is not clearly defined, that is there is no immediate user of their research. Research in an academic world is conducted by doctoral students, post doctorate fellows and research faculty. If the research output does not have immediate application, they take satisfaction of getting a nice publication or two and move on. If the research finds an application few years down the road, the user picks it up, may take advantage of this past work and rip the benefit by commercialization. It is advantageous for universities to bring a likely end user of the research as a partner early in who will support the development through commercialization phase. In recent years some universities have formed a separate office of liaison for getting university research out to the industry.

SCREENING AND PRIORITIZATION

In a prolific R&D environment, testing a new idea is often more difficult than it would appear. Because many ideas are generated on a regular basis, they compete with one another for resources, whether monetary or manpower, or as simple as lab space or time allocation. Project managers must judiciously and impartially select ideas for testing and evaluation according to perceived value, that is, long-term economic return or short-term economic solutions. Screening ideas and prioritizing is a critical step and focus always should on identifying the right thing to do.

An idea selection process is fraught with difficulties and pitfalls and, more painfully, with bruised egos. Project selection is often based on a "quick and dirty" market and economic analysis, which, at the early stages, is almost by definition woefully inadequate. Unfortunately, few other ways exist for passing early judgment on new ideas or principles. The best scheme is to allow as many ideas as possible to be tested up to a preliminary proof-of-principle stage, which means that just enough research is done to determine viability and provide a preliminary assessment of potential.

ESTABLISHING PROOF OF PRINCIPLE

In simple terms, establishing "proof of principle" means determining that the technology has a working chance of success from a technical point of view, and it involves some simple demonstration at the laboratory or in a small bench scale pilot lab unit. This is a critical stage that requires "out-of-the-box" thinkers – individuals with creative and innovative characteristics. Quite often, working on one type of chemistry or catalyst may lead to a different project and a new R&D program. One needs to keep thinking of long range financial goals and opportunities that may arise from some unexpected results. To just cite one example: during the early 1960s UOP developed a process for the separation of linear paraffins from a gasoline range hydrocarbon mixture (isoparaffins, naphthenes, aromatics) so that the linear paraffins could be isomerized and put back into the gasoline, thus improving its octane. Because the value of the incremental octane benefit was low, the process was considered uneconomical and was not commercialized. A few years later, in the mid-1960s there was a need for C_{10}-C_{14} range linear paraffins for manufacturing linear alkylbenzene (LAB), a raw material for biodegradable detergents. As many of you know, the UOP Molex™ process has been a great success that now provides linear paraffin raw material for the manufacture of LAB in every continent. During the 1970s, as a result of the lead phase-out from gasoline, demand for higher-octane gasoline production came and the original concept was finally commercialized for gasoline range application. It may be seen from this example that the commercial development of a technology may take a circuitous route that may diverge from its original purpose.

MARKET AND BUSINESS ANALYSIS

Knowledge of the *market* is essential. Sometimes a market can be created where none existed. One must understand the potential of the innovation. In 1980s no one would have predicted the market potential of cell phones or the Internet as it exists today. An economic analysis can provide us with a rough view of the market potential by counterbalancing all the revenue and income streams against all the R&D cost elements. The net result, after assessing probability and discount factors to reflect risk and delayed time effects, is reflected in a *net present value* (NPV). At this stage, information is the most-valuable commodity. Not just present-day information is needed, but rather the ability to foresee the multiple future market scenarios that may develop. Not all past predictions have been accurate.

- "This 'telephone" has too many shortcomings to be seriously considered as a means of communication" Western Union Memo-1876

- "Heavier-than-air flying machines are impossible" Lord Kelvin, President of the Royal Society-1895

- "The wireless music box (radio) has no imaginable commercial value" David Sarnoffs Associates in response to others' urgings for investments in the radio-1920

- "I think there's a world market for maybe five computers" Thomas Watson, Chairman IBM-1943

- "Computers in the future may weigh no more than 1.5 tons" Popular Mechanics forecasting the relentless march of science-1949

- "There is no reason anyone would want a computer in their home" Ken Olson, President, Chairman and Founder of Digital Equipment-1977

An R&D project may go through a 5-year or longer period of gestation before the birth of the resulting technology. An understanding of the current market and the key market scenarios for the future helps to make a best guess as to what will happen when the new technology hits the market. Similar competing technologies already may be available. Are we well prepared or do we have contingency plans for this event? Will the end product market grow or stagnate or even collapse and disappear? Research and development is a business, and R&D project selection and prioritization procedures must be conducted as such. Management faces a challenge because there is a high degree of uncertainty in the factors required for a thorough business analysis. The team leader needs to keep weaving the pieces of the puzzle into a continuous process as it goes along.

Phase 2: PROCESS DEVELOPMENT

Process development consists of iterative steps. The process begins with the establishment of a proof of principle to assess the validity of the idea and, as noted previously, to allow for at least a preliminary business analysis. Once the R&D project is established, the first experiments are conducted on a small bench scale. The expressed purpose may be to screen various catalyst formulations or determine the range of operating parameters. The statistical design of experiments can and should play an important role at this and later stages as a way to better understand the results and minimize redundant efforts. Because a bench-scale "pilot plant" does not often have product recovery or internal recycle streams built in, the results need to be taken with a certain level of healthy skepticism. **The results achieved under a perfectly controlled environment may not translate as well, or not at all, to real-life situations.**

Once the bench scale laboratory stage is complete, a pilot plant must be designed, constructed, and operated. In this pilot plant, the impact of all the various recycle streams and impurity buildups must be fully assessed. Deactivation of the catalysts or adsorbents needs to be determined over extended periods. All information necessary for a detailed engineering design is collected. **Although a team leader is leading and managing the project, it is the cross-functional team that is setting the design of experiments.** Though the commercial design of the project may be several years away, it is essential that process design and detail mechanical design engineers participate at this early stage to identify the information they will need for the design of a commercial unit. Input from all the various branches of engineering design is critical at this early stage. Past experience is very valuable at this stage. **In an academic environment, it is difficult to form such experienced cross-functional team.**

Phase 3 - DEMONSTRATION UNIT:

The philosophy of various R&D organizations can differ vastly at this point. Some companies insist in building large semi-commercial demonstration units. Other companies run progressively through larger and larger pilot plants, a long, expensive, and time-consuming process. UOP on the contrary has built a well-earned reputation for its ability to scale up from essentially a bench scale unit all the way to full commercial scale. However, this is changing with maturity of the industry. During the early stages, at the birth of the new product introduction, the relative demand is lower and the demonstration size unit could be the first commercial unit. As industry matures, product utilization has grown, more producers, competitors have also entered the market, a first economic size plant of replacement technology for the same product at this stage will require large production capacity. For example, first catalytic reformer unit commercialized in 1949 had capacity of 1500 barrels per day. Today new reformer units are typically 50,000 barrels per day or larger. In a mature competitive industrial atmosphere, producers are less inclined to take a risk to try a radically new technology for the same product. Demonstration phase is very expensive, such demonstration plants may cost anywhere between 10 to 100 million US dollars. This is another reason universities needs to establish a partner, user of the technology.

SCALE-UP AND COMMERCIAL DESIGN

A well-planned R&D project requires a concerted engineering effort at various levels. A conceptual engineering design that reveals what physical or chemical parameters need to be established in the course of the development program and what, if any, new engineering principles needed are developed. Parallel engineering design efforts during the operation of the pilot plant that will provide a semi-complete engineering design, process flow diagrams, and equipment sizes and specifications followed by sufficient engineering work to develop a cost estimate of an actual commercial plant is done to assist ongoing marketing/economic analysis. A complete basic engineering design package, preferably with a complete P&ID, which can provide a full investment cost estimate and engineering analysis of the entire process plant with full safety and hazard analysis including start-up, shut-down, and emergency procedures. Some of these steps may have to be repeated as often as necessary if the information available is proven to be incomplete. This well coordinated engineering effort from the early development phase is critical to the successful implementation of an actual commercial project. The scale-up process requires that the R&D project leaders and the associated engineering personnel work together in a team effort.

ECONOMICS, MARKET, AND BUSINESS ANALYSIS LOOP

Before a new product is launched, or in this case before a new process technology is introduced in the market, a careful and detailed market and business analysis is performed. This analysis was already performed at the early stages of the R&D project but, in reality, this analysis must be frequently updated throughout the project. New relevant information, such as changes in R&D results, process design, market trends, and competitive position, is added as appropriate. The net result is that before the new technology is ready for market introduction, a fully detailed overview of its market and business position must be available.

Marketing comprises the study and analysis of the marketplace, or that part of the world in which the products are to be sold. Marketing requires an understanding of such factors as industry trends, value drivers for different industry segments, demand and supply balances, growth trends and potential, competitive and substitution pressures, and pricing strategies. Marketing also requires a clear understanding of the product that is being sold, its strengths and its weaknesses, its economics, its environmental impact, its applications, its quality or purity specifications, and how it fulfills the customer needs. In summary, marketing is the process whereby a product is positioned in the market.

Phase 4 - COMMERCIALIZATION

The commercialization of a new technology offers untold challenges and opportunities. New technologies present tremendous opportunities to provide new value to customers and to carve out market share. At times, new technologies can even create new markets. If the technology is to be used in-house, the implementation of a commercial project based on that technology requires a financial analysis and the formation of the internal infrastructure necessary for implementation.

If a technology is developed for the purpose of licensing it to others, the company must face the challenge of securing a first customer for that first commercial project. People are naturally conservative, and often understandably so, and are sometimes reluctant to invest in an unknown. The best candidates for commercialization have to be identified early in the marketing process. Therefore, even before the entire technology development process is complete, prospective customers are approached with the most-relevant facts concerning the new technology.

PROJECT MANAGEMENT IN AN R&D ENVIRONMENT

The previous sections have discussed a sequential approach to technology development that is often not a straightforward procedure but rather may take a tortuous path. Undertakings by R&D are expensive; they can all too easily waste considerable amounts of talent, money, and manpower if not properly managed. Research and Development needs to be conducted under a strategic umbrella that, not only sets priorities for projects, but also leads the projects to final completion in the shortest possible time and for the least possible cost. Modern tools of stage-gate reviews, six-sigma and project management are crucial to a well-conducted R&D project.

Project management can be used to shorten the cycle time for R&D projects. With this goal in mind, the focus of project management is to identify and find solutions for problems that are causing schedule delays in the technology delivery process. The concepts involved in successful project management are:

Stage-gate - Risk analysis and decision making
Portfolio management
FMEA
Cross-functional Team and Communication

The *stage-gate* approach is a business process that divides development activities, from the initial concept to final commercialization, into separable stages. Each stage incorporates tasks, or deliverables, that must be completed before that stage is completed. The development teams and their overseers, or gatekeepers, must review the status of the project and the quality of the deliverables at given gate intervals between stages before development can proceed to the next stage. Techno-economic analysis, market analysis and risk assessment are revisited at each stage gate review to support the decision either to move forward to next stage or stop the project. The gatekeepers are a team of cross-functional experienced senior managers.

Portfolio management loosely describes a number of techniques used to manage R&D projects. This is more critical and important for a large R&D organization. In an academic world the research projects are limited to specialization of the research facility. Even in this case, such technique to select a project for funding that will have commercial value in a near future. *Decision and risk analysis* is a tool that is closely associated with the implementation of portfolio management. It involves the statistical analysis of inputs concerning technical, marketing, manufacturing, and other risks associated with a particular development effort.

FMEA: Failure Mode and Effects Analysis is simply an approach designed to ensure that the technical specifications of the process technology are consistent with the requirements. Focus groups are an excellent mechanism to test the waters with "what if" questions as a way of positioning new ideas or concepts. This is a very important step for a first-of-a-kind new process design. To assure the safety of the process one must consider each and every conceivable scenario. The final process design incorporates proper instrumentation and on-line analyses to assure product quality as well as to assure a safe environment and the detection of leaks of any hazardous material, if any were to occur. The FMEA tool also lays the foundation for writing an operating manual as well as establishing emergency procedures. It is very important have a management command structure established for handling an emergency situation. One should keep this well in mind to avoid incidents like Bhopal or Chernobyl.

Cross-functional teams are perhaps the most important contribution to a successful R&D project. Nobody in a company or organization should think they live in an ivory tower. For a successful development of a project one needs experts, specialists from several areas. Even within R&D one would need specialists from the various branches of technology. For example, while developing a process for the catalytic dehydrogenation of propane to propylene, the UOP Oleflex™ process, we had a team that included catalysis experts, surface science and characterization specialists, process design engineers, reaction kinetics specialists, heat exchanger specialists, reactor design engineers, pumps and compressors specialists, and so on. An assertive and effective team leader plays a very important role. Above all, a visionary and a champion are critical in keeping the project on track over the long torturous path.

Communication is paramount to the success of any team effort. All the members of the team should regularly and frequently communicate with one another and also with other colleagues and management. Contacts with other colleagues provide external feelers that at times may point in unexpected directions: teams should always incorporate experts and non-member colleagues as consultants.

MANAGEMENT RESPONSIBILITY

It is critical that R&D as well as the business management stay involved throughout the R&D project. Management provides vision, foresight; establishes goals, and sets up stage gate reviews. It encourages exploration and provides stimulating environment and, most importantly, rewards failures. Yes, it rewards failures, as management must understand that failures lead to success. If in an R&D environment there are no failures it means that the researchers are too cautious, are risk averse, or do not have creative or innovative spirit to try new things. Successes come from those who fail and keep trying again and again.

> ***I have not failed 700 times, I have succeeded in proving that those 700 ways will not work. When I have eliminated the ways that will not work, I will find the way that will work.***
>
> ***Thomas Edison***

Another important function that management must play is to continuously reinvest in improving the R&D infrastructure and the modernization of tools. In today's competitive world, maintaining an efficient R&D organization is essential. Management should invest in new techniques: automated multi-bank micro reactors, combinatorial chemistry, novel tools for determining microstructures, atomic resolution thermal imaging, and so forth.

CONCLUSIONS

Process technology development and commercialization is a serious business that can provide substantial rewards. However, the risks cannot be underestimated. Research and development resources need to be devoted to continuously improving existing technologies to keep them at the forefront and ahead of the competition. Some of the critical success factors include:

- Understanding of critical business issues
- Top management leadership
- Design by cross-functional teams of people working inside the process
- Attention to change management
- A champion
- Efficient project management

The R&D efforts affect all levels of management, from those who allocate resources to those who execute programs. More important, however, R&D resources need to be allocated to the development and commercialization of novel technologies that allow for discrete and substantial added value to customers rather than just only modest enhancements; this part of R&D requires the greatest risk exposure but ultimately may lead to the largest added value. In the process, there will be failures and disappointments. What is important is that the R&D organization perseveres in its resolve to be best in its class and to maintain the morale and high standards – reward the failures and do not punish.

Resources for Inventors, Industry, and Start-Ups

Justin Anderson
Wisconsin Alumni Research Foundation
March 15, 2011

Topical 5: University Research for Industry
Abstract ID# 212492

WARF Wisconsin Alumni Research Foundation

Inventor Resource Overview

- Who can disclose to WARF?
- Am I obligated to disclose my invention to WARF?
  - University policy?
  - Obligations under Bayh-Dole?
  - Sponsored research agreements?
- Why disclose/patent?
  - Costs?
  - Benefits?
- Who should I contact?
  - Staff experience?
  - Scientific expertise and research field?
- How do I disclose?

WARF Wisconsin Alumni Research Foundation

Protecting Your Idea

- Information and materials relating to:
  - Lab notebooks
    - What to record
    - How to record
  - Confidentiality agreements
    - Description and template
  - Material transfer agreements
    - Questions to determine appropriate MTA
    - Templates
  - Publications and other public disclosures
    - Examples
    - Implications

How Do I Disclose?

- Invention Disclosure Report (IDR)
  - Hard copy and/or electronic version
  - Critical information
    - Description of invention
    - Comparable technologies
    - Benefits over existing solutions
    - Funding information
    - Inventors
    - Public disclosures
    - Companies that may be interested in your invention
- In-Person Meeting
  - Questions and discussion
- Website outlines process and provides an FAQ
- IDR can be submitted via website

WARF Wisconsin Alumni Research Foundation

IP Basics and Process

- Website provides both high-level and detailed descriptions of the following:
  - Types of intellectual property
    - Patents
    - Copyrights
    - Trademarks
  - Working with a patent attorney
  - Application contents
    - Specification, figures, claims
  - Patent Office activities
    - i.e. prosecution
  - Timing and patent term

Licensing – For Inventors

- Who will manage my technology?
  - Dedicated IP manager and Licensing manager
  - Each with industry experience
- How can I help?
  - Educate your management team
  - List potential licensees
  - Identify commercial advantages
  - Be available to answer questions
- When does licensing begin?
- What are the common strategies?
  - Direct contacts
  - Conferences, trade shows, technical presentations
  - Website

WARF Wisconsin Alumni Research Foundation

Licensing – For Potential Licensees

- Technology listings
  - Patents and applications
  - Overview
  - Invention specifics
  - Potential uses
  - Key benefits
  - Inventor information
  - Related publications
- Technology updates
  - RSS feed
  - Email based on areas of interest
- Contact forms and directories

WARF Wisconsin Alumni Research Foundation

Licensing – The Process

- Requirements
  - Recognize societal and/or commercial benefit
  - Capability to develop
  - Commitment to commercialize
  - Able to pay a reasonable licensing fee
    - Upfront, royalties, milestones, etc.
- Agreements
  - Descriptions
  - Typical elements
  - Templates

WARF Wisconsin Alumni Research Foundation

Licensing – Templates

- Commercial license
  - Exclusive
  - Exclusive with the right to sublicense
  - Non-exclusive
- Option
- Research license and option
- Standstill
- Material transfer agreement (MTA)
- Inter-institutional agreement (IIA)
- Confidentiality agreement (CDA)
- Print and Sign
  - Limited to a select subset of technologies

Startups

- Assistance for faculty and staff for WARF licensed technology
- Process
  - Before meeting with WARF
    - Feasibility
  - Contact WARF
    - Standstill period
    - Business plan
    - Application form
  - Follow-up actions
    - Agreements, terms, etc.
- Additional Resources
  - Investment and funding
  - Connections: mentors, business community

WARF Wisconsin Alumni Research Foundation

Additional Tools

- Information for inventors
  - Patent searching
  - Research funding
  - Early stage technology development
  - Technology transfer (AUTM)
- Information for licensees
  - UW-Madison Office of Corporate Relations
  - UW-Madison Office of Research and Sponsored Programs
- For both
  - Market research

WARF Wisconsin Alumni Research Foundation

COUPLED ELECTROSTATIC AND PNEUMATIC MICROVALVES FOR PORTABLE CHEMICAL MICROSYSTEMS

Joshua D. Tice, University of Illinois, Urbana, IL
Amit V. Desai, University of Illinois, Urbana, IL
Thomas A. Bassett, University of Illinois, Urbana, IL
Christopher A. Apblett, Sandia National Laboratories, Albuquerque, NM
Paul J. A. Kenis, University of Illinois, Urbana IL

Abstract

In order to produce elaborate microchemical systems that are both highly integrated and portable, one crucial enabling technology that needs to be further developed is the microvalve. Pneumatic microvalves have been widely implemented in academic studies[1] and also manufactured as part of commercially available microfluidic systems.[2] However, the portability of such systems is limited by the bulk of ancillary equipment, including an external source of pressure, an array of solenoid valves, and computerized controls. In order to retain the versatility of pneumatic valves while improving their portability, we have developed a microvalve that is actuated with electrostatic forces and can be used to direct pressure to the control lines of pneumatic valves. With a compact source of pressure, these electrostatic valves circumvent the scaling issues encountered with solenoid valve arrays. The presentation will include a description of the current fabrication process – accomplished under ambient pressure and moderate temperatures using soft-lithographic techniques – followed by a discussion of envisioned routes to manufacturing. Additionally, characterization of the valve's actuation potentials and isolation pressures will be presented. With further development, the valves have the potential to address needs for miniaturized point-of-care diagnostics or portable chemical sensing, among other applications.

References

1. Unger, Marc A.; Chou, Hou-Pu; Thorsen, Todd; Scherer, Axel and Quake, Stephen R. (2000), "Monolithic Microfabricated Valves and Pumps by Multilayer Soft Lithography," *Science*, vol. 288, pp. 113-116.
2. www.fluidigm.com, accessed January 24, 2011.

(66e) Normally Closed Valves: Design Considerations, Scope and Applications

Ritika Mohan[1], Paul J. A. Kenis[2], Amit V. Desai[3] and Benjamin R. Schudel[2], (1)University of Illinois, Urbana Champaign, Urbana, IL, (2)Department of Chemical & Biomolecular Engineering, University of Illinois at Urbana-Champaign, Urbana, IL, (3)Chemical & Biomolecular Engineering, University of Illinois at Urbana-Champaign, Urbana, IL

Microvalves are one of the most important components of microfluidic devices. Though a lot significant progress has been made in improving the performance of these valves with respect to their size, response time, leakage, dead volume, power consumption, chemical compatibility, and sensitivity to particulate contamination, improvements need to be done for a better functionality of microfluidic device.

Incorporation of all the ideal characteristics in a microvalve is challenging, however, depending on the application, a microvalve can be designed to have most of the desired properties. For instance, pneumatic microvalves developed by Quake and coworkers have been successfully used in many applications requiring multi-step and high throughput operations on a single device [1]. One of the biggest advantage is the compatibility of these valves with standard soft lithography which allows easy integration into complex microfluidic devices.

Pneumatic microvalves can be classified into two types of valves: normally open and normally closed. Although normally open valves are widely used in many microfluidic applications, they have limited portability in applications that require continuous closed state for long periods of time, as these microvalves need bulky ancillaries for actuation. For example, in investigation of protein-antibody interaction, the valve needs to be open only for a short period of time, when the solutions are being mixed. In such applications, since normally open valves require continuous pump actuation, they are limited in their portability between filling section and detection ancillaries [2]. Another limitation of normally open microvalves are their high actuation pressures, up to 28 psi, especially for taller channels. Normally closed valves not only address the above limitations, but also retains the ease of fabrication and integration on to microfluidic devices.

Our focus is on the design rules of these microvalves because actuation of these microvalves involves complex interplay between mechanical deformation and adhesion which cannot be quantified easily. In addition, design rules become even more important in devices with dense network because pressure losses across the device cause failure of the valves at the end of the network and hence, minimizing actuation pressures becomes crucial. Finally, since they are compatible with standard soft lithography, optimization for their integration into complex microfluidic devices is necessary.

References:

[1]Schudel, Benjamin R, et al. "Microfluidic chip for combinatorial mixing and screening of assays." Lab on a Chip (2009): 1676-1680.

[2]Unger, Marc A, et al. "Monolithic Microfabricated Valves and Pumps by Multilayer Soft Lithography." Science (2000): 113-116.

See more of this Session: Industry Needs and Goals for University Research
See more of this Group/Topical: Topical 5: University Research for Industry

(66f) Thermodynamic Data, Physicochemical Properties, and Molecular Information on More Than 2.5 Million Chemical Compounds

Yangsoo Kim, Won-chon Jung, Ae-ri Sung, Jeongjae Jeon, Yunkyung Kwon, Junhyuk Cho, O Hyoung Kwon, Yongkun Leem and **Tae-Yun Park**, Research & Development, Visual Simulation, Co. Ltd., Seoul, South Korea

Scientists and engineers often suffer from the shortage of thermodynamic data, physicochemical properties, and molecular information of chemical compounds. This is not surprising if one consider the fact that over 50 million chemicals have been registered as of October 2010, while the number of chemicals whose thermodynamic or physicochemical data are available is only in the order of 10,000. Experimental measurements are expensive and time consuming, and reliable calculation methods are rarely available.

We have developed computer modules for the prediction of thermodynamic data and physicochemical properties based on quantum mechanical calculations at decent accuracy levels and more than 2,000 molecular descriptors. The predicted results were carefully verified with millions of experimental data collected for over 3 years by more than 50 scientists and engineers, which confirmed that our computer modules are capable of predicting thermodynamic data and physicochemical properties at the accuracy level of experiments.

An automatic procedure from the molecule generation to the prediction of the data and information has been developed, and a computing center containing over 550 computers has been constructed to process massive amount of chemical compounds. Using the procedure and the computing system, more than 2.5 million chemicals have been processed. The final data and the information have been packed into a database server. To search and browse the data of the target chemical, information browsing software has also been developed, which provides the access to the database server and information retrieval online.

Our database contains a total of 2,140 data and information sets per molecule, consisting of 46 thermodynamic data and physicochemical properties, 3 spectra (IR, NMR, VCD), 69 quantum mechanical information, 2,004 molecular descriptors, and 18 drug related properties. The 2.5 million chemical compounds contains radicals, hydrocarbons, fuels such as gasoline, jet-fuel, diesel, bio-diesel, etc., compounds involved in the commercial processes like thermal cracking, combustion, reforming, isomerization, and drug-like molecules and hetero-compounds containing oxygen, nitrogen.

Due to the limitation of quantum computation time, molecules containing other than C, H, N, O, and S atoms and the compounds with number of carbon atom above C_{25} were not processed. They will also be processed in the near future after upgrading our current computing system.

See more of this Session: Industry Needs and Goals for University Research
See more of this Group/Topical: Topical 5: University Research for Industry

AUTHOR INDEX